PERFORMANCE EVALUATION OF BEARINGS

Edited by **Rakesh Sehgal**

Performance Evaluation of Bearings
http://dx.doi.org/10.5772/2421
Edited by Rakesh Sehgal

Published by InTech

Janeza Trdine 9, 51000 Rijeka, Croatia

Publishing Process Manager Mirna Cvijic

Technical Editor Teodora Smiljanic

Cover Designer InTech Design Team

Additional hard copies can be obtained from orders@intechopen.com

Performance Evaluation of Bearings, Edited by Rakesh Sehgal
p. cm.
ISBN 978-953-51-0786-6

Meet the editor

Rakesh Sehgal is currently Professor at the Department of Mechanical Engineering and Dean (Faculty Welfare) at National Institute of Technology, Hamirpur, Himachal Pradesh, India. He did his post graduation in Design of Mechanical Equipment from Indian Institute of Technology, Delhi and Doctorate in Tribology from Kurukshetra University. Currently, he is actively engaged in research in tribology and pursuing post doctorate in the area of non-circular journal bearings under PDF Award from University Grants Commission, New Delhi, India. He has supervised six doctorate students and several M.Tech. students. He has completed several sponsored research projects and published more than 104 research papers in peer reviewed international/national journals and conferences. He is a life member of Institution of Engineers (India), Indian Society for Technical Education (ISTE), Association for Machines and Mechanisms (AMM), Indian Society of Theoretical and Applied Mechanics (ISTAM), Tribology Society of India (TSI), Materials Research Society of India (MRSI) and Member of ASME.

Contents

Preface

Bearings (both plain and rolling element) are used as important supporting elements for locating rotating components and confining their motion in desired direction. In order to ensure their operational reliability and desired life, these need to be properly designed/selected for an application more so because of ever increasing operational speeds. This requires a careful performance evaluation of different types of bearings considering aspects such as thermal stability, lubrication, contaminants in lubricants and controlling mechanism etc.

Traditionally, researchers have been focusing on both experimental as well as theoretical studies for the design and performance evaluation of bearings. However, theoretical studies have so far dominated the scenario with development of modeling and computational techniques like finite difference, finite element, fuzzy logic and neural networks and artificial intelligence. Apart from these techniques, scientists have the freedom to use different standard modeling and analysis tools developed from time to time for design and evaluation of bearings.

The title of this book was specifically chosen as ***Performance Evaluation of Bearings.*** The present book is a compilation of different aspects contributing towards the performance evaluation of plain bearings (both journal and thrust), rolling element bearings and magnetic bearings. The book is divided into three broad sections and eight chapters presenting both theoretical and experimental work. Chapter 1 covers plain bearings, chapters 2 - 5 deal with rolling element bearings and chapters 6 - 8 present the treatment of magnetic bearings. The work presented in the book is quite relevant and useful to researchers working/intend to work in the area of design and performance evaluation of bearings both fundamental and advanced.

Dr. Rakesh Sehgal,
National Institute of Technology,
Hamirpur
India

Section 1

Plain Bearings

Chapter 1

Thermal Studies of Non-Circular Journal Bearing Profiles: Offset-Halves and Elliptical

Amit Chauhan and Rakesh Sehgal

Additional information is available at the end of the chapter

http://dx.doi.org/10.5772/51261

1. Introduction

Hydrodynamic journal bearings are defined as the mechanical components that support the external loads smoothly due to geometry and relative motion of mating surfaces in the presence of a thick film of lubricant. Hydrodynamic journal bearings are extensively used in high speed rotating machines because of their low friction, high load capacity, and good damping characteristics. Such bearings have many different designs to compensate for differing load requirements, machine speeds, cost, and dynamic properties. One unique disadvantage which consumes much time towards the research and experimentation is an instability which manifests itself as oil whip which is a vibration phenomenon. Oil whip is disastrous because the rotor cannot form a stable wedge and consequently this leads to metal to metal contact between the rotor and the bearing surface. Once surface contact exists the rotors begins to precess, in a reverse direction from the actual rotor rotation direction, using the entire bearing clearance. This condition leads to high friction levels which will overheat the bearing metal thus causing rapid destruction of the bearing, rotor journal and machine seals. Fuller [1956] has suggested that the fluid film bearings are probably the most important mechanical components in the recent technological development and are comparable in their significance to the effect of electricity. The development of fluid film lubrication mechanisms has been observed by Petrov [1883] in Russia and Tower [1883] in England. In 1886, Reynolds presented his classical analysis of bearing hydrodynamics, which forms the basis of present days bearing study. The overview of both the circular and non-circular hydrodynamic journal bearings and their design methodologies are discussed as follows:

1.1. Circular journal bearing

The basic configuration of the circular journal bearing consists of a journal which rotates relative to the bearing which is also known as bush (Fig. 1). Efficient operation of such bearing requires the presence of a lubricant in the clearance space between the journal and the bush. In hydrodynamic lubrication it is assumed that the fluid does not slip at the interface with the bearing and journal surface i.e. the fluid in contact with the journal surface moves at the same speed as the journal surface. Over the thickness of the fluid there is a velocity gradient depending on the relative movement of the bearing surfaces. If the bearing surfaces are parallel or concentric, the motion of the lubricant will not result in pressure generation which could support bearing load.

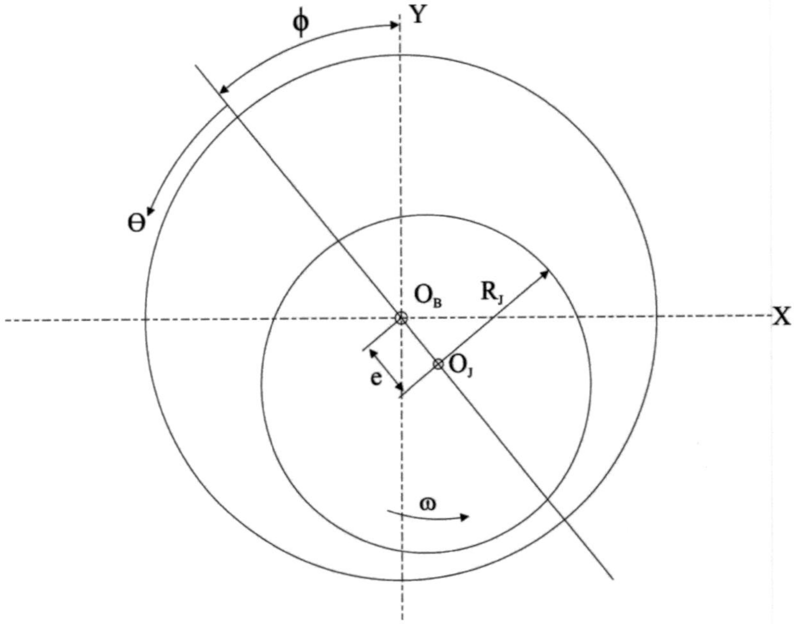

Figure 1. Schematic of circular journal bearing

However, if the surfaces are at a slight angle, the resulting lubrication fluid velocity gradients will be such that generation of pressure results from the wedging action of the bearing surfaces. Hydrodynamic lubrication depends upon this effect. The operation of hydrodynamic lubrication in journal bearings has been illustrated in Fig. 2. Before the rotation commences i.e. at rest the shaft rests on the bearing surface. When the journal starts to rotate, it will climb the bearing surface gradually as the speed is further increased; it will then force the lubricant into the wedge-shaped region. When more and more lubricant is forced into a wedge-shaped clearance space, the shaft moves up the bore until an equilibrium condition is reached and now, the shaft is supported on a wedge of lubricant. The moving surfaces are then held apart by the pressure generated within the fluid film. Journal bearings are designed such that at normal operating conditions the continuously generated fluid pressure supports the load with no contact between the bearing surfaces. This operating condition is known as thick film or fluid film lubrication and results in a very low operating friction. On the other hand if the lubricant film is insufficient between the relatively moving parts, it may lead to surface contact and the phenomenon is normally known as boundary lubrication. This occurs at rotation start-up, a slow speed operation or if the load is too heavy. This regime results in bearing wear and a relatively high friction value. If a bearing is to be operated under boundary lubricating conditions, special lubricants must be used. Amongst hydrodynamic bearings, circular journal bearing is the most familiar and widely used bearing. Simple form of this bearing offers many advantages in its manufacturing as well as in its performance. However, the circular journal bearings operating at high speed encounter instability problems of whirl and whip. Instability may damage not only the bearings but also the complete machine.

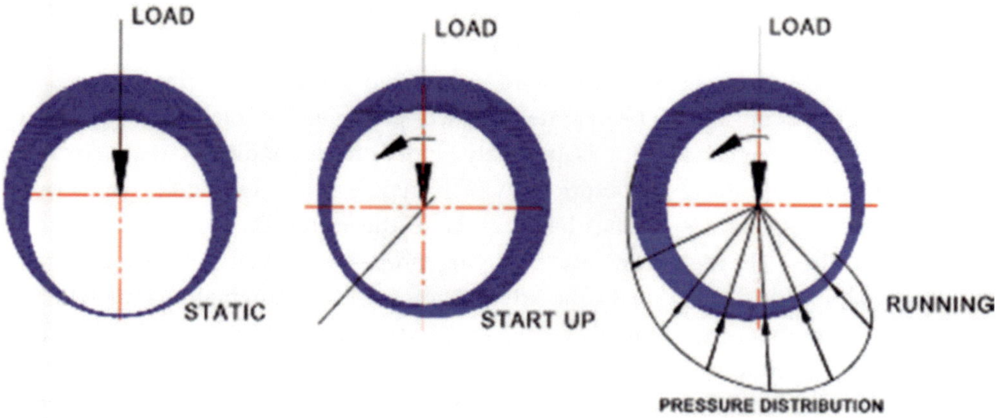

Figure 2. Schematic of operation of hydrodynamic lubrication in journal bearing [W1]

Moreover, these bearings usually experience a considerable variation in temperature due to viscous heat dissipation. This significantly affects the bearing performance as lubricant viscosity is a strong function of temperature. Furthermore, excessive rise in temperature can cause oxidation of the lubricant and, consequently, lead to failure of the bearing. Pressure also influences the viscosity of the lubricant to certain extent. Usually viscosity increases exponentially as the pressure increases which in turn increases the load capacity of the journal bearing. Researchers have studied the behaviour of circular journal bearing by adopting various numerical approaches to simulate the performance in accordance with the real conditions.

1.2. Non-circular journal bearing

It has been reported in the literature that the temperature rise is quite high in circular journal bearings as they operate with single active oil film. This resulted in the development of bearings with non-circular profile which operate with more than one active oil film. This feature accounts for the superior stiffness, damping, and reduced temperature in the oil film as compared to the circular journal bearings. Almost all the non-circular journal bearing geometries enhance the shaft stability and under proper conditions this will also reduce power losses and increase oil flow (as compared to an inscribed circular bearing), thus reducing the oil film temperature. Amongst non-circular journal bearings, offset-halves, elliptical, lemon bore, and three-lobe configurations are the most common.

The offset-halves journal bearing has been commonly used as a lobed bearing in which two lobes are obtained by orthogonally displacing the two halves of a cylindrical bearing. Offset-halves journal bearings (Fig. 3) are frequently used in gear boxes connecting turbine and generator for the power generation industries. These also find applications where primary directions of force, constant direction of rotation are found or high bearing load capacity, long service life, high stiffness, and damping values are the main characteristics under consideration. If the equipment is operated at full power, these requirements can be met by lemon bore bearings. Lemon bore bearing is a variation on the plain bearing where the bearing clearance is reduced in one direction and this bearing has a lower load carrying capacity than the plain bearings, but is more susceptible to oil whirl at high speeds [W2]. However, equipment must often be operated at lower performance levels, particularly in the times of reduced current needs. It is precisely under these conditions that lemon bore bearings may provide unstable conditions, which may require equipment shut down to avoid damage. Offset-halves journal bearings have the durability equal to lemon bore bearings while these show stiffness and damping properties which permit light loads at high rotational speeds. It also offers the advantage of a long, minimally convergent inlet gap, resulting in high load carrying capacity. At the same time, the externally applied force and the compression resulting from the horizontal displacement of the bearing halves accurately holds the shaft in the lubricant film. This effect produces excellent hydrodynamic characteristics, such as elastic rigidity and damping by the oil film. Thus, the offset-halves journal bearings prove to be technical alternative to conventional lemon bore bearings [Chauhan and Sehgal: 2008].

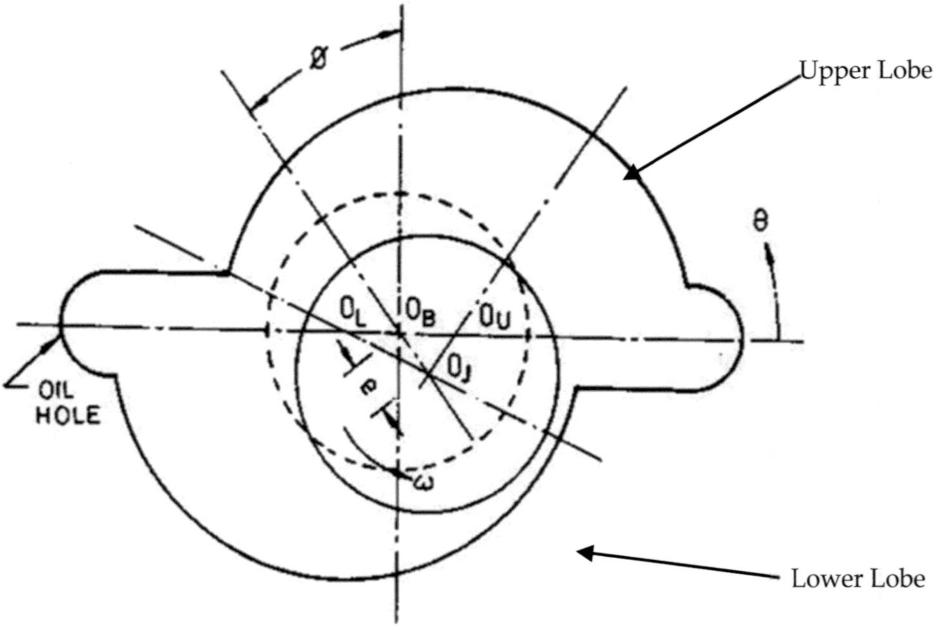

Figure 3. Schematic diagram of offset-halves journal bearing

The elliptical journal bearings (Fig. 4) are commonly used in turbo-sets of small and medium ratings, steam turbines, and generators. The so-called elliptical journal bearing is actually not elliptic in cross section but is usually made up of two circular arcs whose centers are displaced along a common vertical straight line from the centre of the bearing. The bearing so produced has a large clearance in the horizontal or split direction and a smaller clearance in the vertical direction. Elliptical journal bearings are slightly more stable toward the oil whip than the cylindrical bearings. In addition to this elliptical journal bearing runs cooler than a cylindrical bearing because of the larger horizontal clearance for the same vertical clearance.

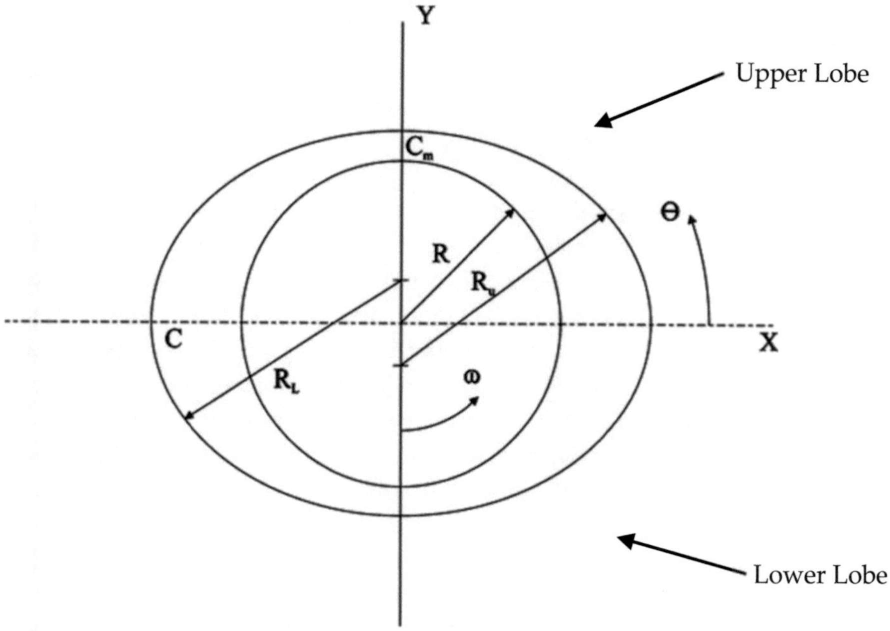

Figure 4. Schematic diagram of elliptical journal bearing

2. Literature review

This section of the chapter provides details of research carried out on hydrodynamic bearings in general and, offset-halves and elliptical journal bearings in particular. There is enormous information available on the theoretical and experimental work related to the circular journal bearings. However, such work pertaining to non-circular journal bearings especially offset-halves and elliptical journal bearings are limited and hence, the theoretical and experimental works pertaining to non-circular journal bearings have been summarized:

Pinkus and Lynn [1956] have presented an analysis of elliptical journal bearings based on the numerical solution of Reynolds equation for finite bearings with the assumption that there is no heat loss to the surroundings. They have supplemented the solution of differential equation by additional work on the nature of the oil flow, power loss, and eccentricity in elliptical journal bearings. Wilcock [1961] has worked towards the possibility of displacing the lobe centers of two-lobe journal bearings orthogonally with respect to the mid-radius of the lobe. The author shows that when the lobe displacement is in a direction opposite to the shaft surface motion, and the bearing is centrally loaded, shaft stiffness orthogonal to the load vector is substantially increased. At the same time, vertical stiffness essentially remains unchanged and minimum film thickness is decreased; particularly at low loads, while oil flow is increased. Author also carried out an analysis for a bearing having in cross-section two arcs (each subtending an angle of 150^0), L/D=1/2, and with the arc centers each displaced from the geometric center by half the radial clearance. Singh et al. [1977] have reported that non-circular bearings are finding extensive use in high speed machinery as they enhance shaft stability, reduce power losses and increase oil flow (as compared to circular bearings), thus reducing bearing temperature. The authors had presented a solution to analyze the elliptical bearings, using a variational approach. Crosby [1980] has solved full journal bearing of finite length for thermohydrodynamic case in which the energy transmitted by conduction is included. The effect of temperature variation across the film thickness on bearing performance is investigated by the author. Singh and Gupta [1982] have considered the stability limits of elliptical journal bearings supporting flexible rotors. The Reynolds equation was solved numerically for several values of the eccentricity ratio, the L/D ratio and the dimensionless velocity of the journal centre. The authors observed that the operating load, ellipticity, L/D ratio and shaft flexibility significantly affect the limit of stable operation. The authors also reported that elliptical bearings are suitable for stiff and moderately flexible rotors. Tayal et al. [1982] have investigated the effect of non-linear behavior of additive-fortified lubricants which behave as non-Newtonian fluid on the performance characteristics of the finite width elliptical journal bearing. The finite element method with Galerkin's technique was used to solve the Navier-Stokes equations in cylindrical co-ordinates that represent the flow field in the clearance space of a bearing using Newtonian fluids, and then the non-Newtonian effect was introduced by modifying the viscosity term for the model in the iterations. Booker and Chandra et al. [1983] have compared the performance of different bearing configurations namely offset-halves, lemon-bore, three-lobe and four-lobe bearing at the same load capacity and speed. During the comparison, the authors have considered the steady state and stability characteristics. Govindachar [1984] have suggested that Novel 'offset' designs offer attractive possibilities in several applications for which conventional journal bearings are only marginally satisfactory. They considered one such prototypical problem in rotating machinery (the support of a rigid rotor turning at high speed under gravity). The problem has been studied by the authors through a numerical example for both dimensional and non-dimensional parametric studies. The authors show that the stability of full journal bearing system is significantly improved by moderate offset and is fairly insensitive to small departures from optimal design values. Singh and Gupta [1984] have theoretically predicted the stability of a hybrid two-lobe bearing which is obtained by displacing the lobe centers of an elliptical bearing. It has been found that an orthogonally displaced bearing is more stable than the conventional bearings. Mehta and Singh [1986] have analytically analyzed the dynamic behaviour of a cylindrical pressure dam bearing in which centers of both halves are displaced. Authors observed that the stabilities of a cylindrical pressure dam bearing can be increased many times by displacing the centers of two halves. It has been reported by the authors that the bearing so obtained is even superior to elliptical and half elliptical pressure dam bearings in stability. Read and Flack [1987] developed a test apparatus on which an offset-halves journal bearing of 70 mm diameter journal was tested at five vertical loads and two rotational speeds. Singh et al. [1989] have studied the effects of linear elastic deformation and lubricant viscosity with pressure and temperature for an elliptical bearing and solved 3-D equations for momentum, continuity and elasticity to obtain pressure in the lubricant flow-field and deformation in the bearing liner. Basri and Gethin [1990] have carried out a theoretical analysis of the thermal behaviour of orthogonally displaced, three-lobe, and four-lobe bearing geometries. The thermal analysis illustrates the implication of the type selection with regards to the parameters of load-carrying ability, power loss, lubricant requirements, and operating temperatures. The comparisons presented by the authors show that all profiles considered have inferior load-carrying ability and less extreme thermal effects when compared with the cylindrical geometry along with significantly larger lubricant supply requirements. Hussain et al. [1996] have predicted the temperature distribution in various non-circular journal bearings namely two-lobe, elliptical, and orthogonally displaced. The work reported by them is based on a two-dimensional treatment following Mc Callion's approach in which the Reynolds and energy equations in oil film are decoupled by neglecting all pressure terms in the energy equation. Ma and Taylor [1996] experimentally investigated the thermal behaviour of a two-axial-groove circular bearing and an elliptical bearing, both 110mm in diameter. Both bearings were tested at specific loads upto 4 MPa and rotational frequencies up to 120 Hz. The authors have measured the power losses & flow rates directly, and the detailed temperature information was collected. The results presented by them show that the thermal effects are significant in both bearings. Banwait and Chandrawat [2000] have analyzed the effect of journal misalignment on the thermohydrodynamic performance characteristics of elliptical journal bearing and solved the generalized Reynolds equation for the oil-film pressure distribution. The energy and heat conduction equations are used by the authors for determining the oil-film, bush and journal temperature fields. This work has reported the important features observed in static performance characteristics of thermohydrodynamic analysis of misaligned elliptical bearing along with the isopressure curves, pressure profiles, isotherms and temperature profiles. Sehgal et al. [2000] have presented a comparative analysis of three types of hydrodynamic journal bearing configurations namely, circular, axial groove, and offset-halves. It has been reported by the authors that the offset bearing runs cooler than an equivalent circular bearing with axial grooves. A computer-aided design of hydrodynamic journal bearing is provided considering thermal effects by Singh and Majumdar [2005]. They have solved the Reynolds equation simultaneously along with the energy equation and heat conduction equations in bush and shaft to obtain the steady-state solution. A data bank is generated by the authors that consists of load, friction factor and flow rate for different L/D & eccentricity ratios. Sharma and Pandey [2007] have carried out a thermohydrodynamic lubrication analysis of infinitely wide slider bearing assuming parabolic and Legendre polynomial temperature profile across the film thickness. It was observed that the temperature approximation across the film thickness by Legendre

Polynomial yields more accurate results in comparison to Parabolic Profile approximation. However, the former is algebraically more complex to tackle in comparison to the later; the authors also observed that the film temperatures computed by Parabolic Profile approximation are lower in comparison to Legendre Polynomial approximation. Further, it has been concluded by them that the computational time taken in solution of coupled governing equation with both temperature profile approximation have only marginal difference. Mishra et al. [2007] have considered the non-circularity in bearing bore as elliptical and made a comparison with the circular case to analyse the effect of irregularity. It was reported that, with increasing non-circularity, the pressure gets reduced and temperature rise obtained is less. Chauhan et al. [2010] have presented a thermohydrodynamic study for the analysis of elliptical journal bearing with three different grade oils. The authors have made a comparative study of rise in oil-temperatures, thermal pressures, and load carrying capacity for three commercial grade oils. Rahmatabadi et al. [2010] have considered the non-circular bearing configurations: two, three and four-lobe lubricated with micropolar fluids. The authors have modified the Reynolds equation based on the theory of micropolar fluids and solved the same by using finite element methods. It has been observed by the authors that micropolar lubricants can produce significant enhancement in the static performance characteristics and the effects are more pronounced at larger coupling numbers.

3. Numerical techniques for the thermal analysis of non-circular journal bearing

In earlier works, the bearing performance parameters have been computed by solving the Reynolds equation only. Over the years, many researchers have proposed number of mathematical models. A more realistic thermohydrodynamic (*THD*) model for bearing analysis has been developed which treats the viscosity as a function of both the temperature and pressure. Moreover, it also considers the variation of temperature across the film thickness and through the bounding solids (housing and Journal). The thermohydrodynamic model also presents coupled solutions of governing equations by incorporating appropriate boundary conditions and considering the heat conduction across the bearing surfaces. Even the importance of *THD* studies in hydrodynamic bearings can be justified by looking at the large volumes of research papers that are being published by researchers using various models. The theoretical investigations have been carried out into the performance of hydrodynamic journal bearing by adopting various methods, which are classified in two categories as: First category is the methods which comprise a full numerical treatment of temperature variation across the lubrication film thickness in energy equation using Finite Difference Method (FDM) or Finite Element Method (FEM). Second one is the methods which incorporate polynomial approximation to evaluate the transverse temperature variation in the lubrication film thickness. Both approaches mentioned can be used for the analysis of hydrodynamic bearings and have certain merits. The first approach is relatively accurate at the expense of computational speed and time, whereas the second is relatively fast at the expense of accuracy.

In present chapter, the thermal studies of non-circular journal bearings: offset-halves and elliptical have been presented using thermohydrodynamic approach.

3.1. Film thickness equation

The film thickness equations for offset-halves journal bearing (Fig. 1 (a)) are given as Sehgal et al. [2000]:

$$h = c_m\left[\left(\frac{1+\delta}{2\delta}\right)+\left(\frac{1-\delta}{2\delta}\right)\cos\theta-\varepsilon\sin\left(\phi-\theta\right)\right] \quad (0<\theta<180) \tag{1}$$

$$h = c_m\left[\left(\frac{1+\delta}{2\delta}\right)-\left(\frac{1-\delta}{2\delta}\right)\cos\theta-\varepsilon\sin\left(\phi-\theta\right)\right] \quad (180<\theta<360) \tag{2}$$

In these equations, h represents film thickness for circular journal bearing, C represents radial clearance, ε represents eccentricity ratio, and θ represents angle measured from the horizontal split axis in the direction of rotation. C_m denotes minimum clearance when journal centre is coincident with geometric centre of the bearing, δ denotes offset factor (C_m / C), and ϕ denotes attitude angle.

The film thickness equations for elliptical journal bearing (Fig. 1 (b)) are given as Hussain et al. [1996]:

$$h = c_m\left[1+E_M+\varepsilon_1\cos\left(\theta+\phi-\phi_1\right)\right], \text{ for } 0<\theta<180 \tag{3}$$

$$h = c_m\left[1+E_M+\varepsilon_2\cos\left(\theta+\phi-\phi_2\right)\right], \text{ for } (180<\theta<360) \tag{4}$$

Different parameters used in eqns. (3) & (4) are given as:

$$\varepsilon_1 = \left({E_M}^2+\varepsilon^2-2E_M\varepsilon\cos\phi\right)^{\frac{1}{2}};\ \varepsilon_2 = \left({E_M}^2+\varepsilon^2+2E_M\varepsilon\cos\phi\right)^{\frac{1}{2}}$$

$$\phi_1 = \pi-\tan^{-1}\left(\frac{\varepsilon\sin\phi}{E_M-\varepsilon\cos\phi}\right);\ \phi_2 = \tan^{-1}\left(\frac{\varepsilon\sin\phi}{E_M+\varepsilon\cos\phi}\right);\ E_M = \left(\frac{C_h-C_m}{C_m}\right)$$

In eqns. (3 and 4), h represents film thickness for elliptical journal bearing, E_M represents elliptical Ratio, $\varepsilon_1,\varepsilon_2$ represents eccentricity ratio from 0-180 Deg. (upper lobe) and 180-360 Deg. (lower lobe) respectively, ϕ_1,ϕ_2 represents attitude angles from 0-180 Deg. (upper lobe) and 180-360 Deg. (lower lobe) respectively and C_h represents horizontal clearance for elliptical journal bearing. Film thickness represented by eqns. 1, 3 corresponds to upper lobe whereas eqns. 2, 4 represents film thickness for lower lobe of the.

3.2. Reynolds equation

All the simplifying assumptions necessary for the derivation of the Reynolds equation are listed below Stachowiak and Batchelor [1993]:

1. Body forces are neglected i. e. there are no extra outside fields of forces acting on the fluids.
2. Pressure is constant through the film.
3. No slip at the boundaries as the velocity of the oil layer adjacent to the boundary is the same as that of the boundary.
4. Flow is laminar and viscous.
5. Lubricant behaves as a Newtonian fluid.
6. Inertia and body forces are negligible compared with the pressure and viscous terms.
7. Fluid density is constant. Usually valid for fluids, when there is not much thermal expansion.
8. There is no vertical flow across the film.

For steady-state and incompressible flow, the Reynolds equation is Hussain et al. [1996]:

$$\frac{\partial}{\partial x}\left(\frac{h^3}{\mu}\frac{\partial p}{\partial x}\right)+\frac{\partial}{\partial z}\left(\frac{h^3}{\mu}\frac{\partial p}{\partial z}\right)=6U\frac{\partial h}{\partial x} \tag{5}$$

Here, P represents film pressure, μ represents absolute viscosity of the lubricant, and U represents velocity of journal.

This equation is set into finite differences by using central difference technique. The final form is reproduced here.

$$P(i,j)_{iso} = A1P(i+1,j)_{iso} + A2P(i-1,j)_{iso} + A3P(i,j+1)_{iso} + A4P(i,j-1)_{iso} - A5 \tag{6}$$

$$P(i,j)_{th} = E11P(i+1,j)_{th} + E22P(i-1,j)_{th} + E33P(i,j+1)_{th} + E44P(i,j-1)_{th} - E55 \tag{7}$$

$$\text{Where, } A11=\left[\frac{h^3}{d\theta^2}+\frac{3h^2}{2d\theta}\frac{\partial h}{\partial\theta}\right];\ A22=\left[\frac{h^3}{d\theta^2}-\frac{3h^2}{2d\theta}\frac{\partial h}{\partial\theta}\right];\ A33=\left[\frac{r^2h^3}{dz^2}+\frac{3r^2h^2}{2dz}\frac{\partial h}{\partial z}\right];$$

$$A44=\left[\frac{r^2h^3}{dz^2}-\frac{3r^2h^2}{2dz}\frac{\partial h}{\partial z}\right];A55=\left[6Ur\mu\frac{\partial h}{\partial\theta}\right];A=\left[\frac{2h^3}{d\theta^2}+\frac{2r^2h^2}{dz^2}\right];$$

$$A1=A11/A;A2=A22/A;A3=A33/A;A4=A44/A;A5=A55/A$$

$$A6=\frac{h^3}{\mu};\ A7=\frac{\partial A6}{\partial\theta};\ A8=\frac{\partial A6}{\partial z};\ A9=\frac{2A6}{d\theta^2}+\frac{2r^2A6}{dz^2};F=\frac{A7}{2d\theta};\ G=\frac{A6}{d\theta^2};\ B=\frac{\partial h}{\partial\theta};$$

$$H=\frac{r^2A8}{2dz};\ H1=\frac{r^2A6}{dz^2};\ H2=6UrB\,;E11=\frac{F+G}{A9};\ E22=\frac{-F+G}{A9};$$

$$E33 = \frac{H + H1}{A9};\ E44 = \frac{-H + H1}{A9};\ E55 = \frac{-H2}{A9}$$

Equation (6) assumes viscosity as constant and gives isothermal pressures, whereas eqn. (7) assumes viscosity as a variable quantity and gives thermal pressures. The coefficients appearing in eqn. (6 & 7) are given in Appendix-I. The variation of viscosity with temperature and pressure has been simulated using the following viscosity relation Sharma and Pandey [2007]:

$$\mu = \mu_{ref} e^{\alpha P - \gamma (T - T_0)} \tag{8}$$

In eqn. (8), μ_{ref} represents absolute viscosity of the lubricant at oil inlet temperature, γ represents temperature-viscosity coefficient of lubricant, α represents Barus viscosity-pressure index, T represents lubricating film temperature, and T_o represents oil inlet temperature.

3.3. Energy Equation

The energy equation for steady-state and incompressible flow is given as Sharma and Pandey [2007]:

$$\rho C_P \left(u \frac{\partial T}{\partial x} + w \frac{\partial T}{\partial z} \right) = \frac{\partial}{\partial y} \left(K \frac{\partial T}{\partial y} \right) + \mu \left[\left(\frac{\partial u}{\partial y} \right)^2 + \left(\frac{\partial w}{\partial y} \right)^2 \right] \tag{9}$$

Here, C_P represents specific heat of the lubricating oil, K represents thermal conductivity of the lubricating oil, and u, w represents velocity components in X- and Z-directions. The term on the left hand side in eqn. (9) represents the energy transfer due to convection, and the first, second terms on right hand side of the eqn. (9) represents the energy transfer due to conduction and energy transfer due to dissipation respectively. In eqn. (9), x-axis represents the axis along the circumference of bearing, y-axis represents the axis along the oil film thickness and z-axis represents the axis across the width of bearing. The variation of temperature across the film thickness in equation (9) is approximated by parabolic temperature profile. It is pertinent to add here that the temperature computed by this approach have been reported to be on lower side in comparison to those obtained through Legendre polynomial temperature profile approximation by Sharma and Pandey [2006]. The temperature profile across the film thickness is represented by a second order polynomial as:

$$T = a_1 + a_2 y + a_3 y^2 \tag{10}$$

In order to evaluate the constants appearing in eqn. (10), the following boundary conditions are used:

$$\text{At } y = 0, T = T_L$$

$$\text{At } y = h, T = T_U$$

and

$$T_m = \frac{1}{h}\int_0^h T dy$$

After algebraic manipulations, the equation (10) takes the following form:

$$T = T_L - \left(4T_L + 2T_U - 6T_m\right)\left(\frac{y}{h}\right) + \left(3T_L + 3T_U - 6T_m\right)\left(\frac{y}{h}\right)^2 \tag{11}$$

Where, T_L, T_U and T_m represent temperatures of the lower bounding surface (journal), upper bounding surface (bearing), and mean temperature across the film respectively.

Final form of the energy equation is represented as:

$$\begin{aligned} &6T_L + 6T_U - 12T_m - \frac{\rho C_P h^4}{120K\mu}\frac{\partial P}{\partial x}\left(\frac{\partial T_L}{\partial x} + \frac{\partial T_U}{\partial x} - 12\frac{\partial T_m}{\partial x}\right) - \frac{\rho C_P h^4}{120K\mu}\frac{\partial P}{\partial z} \\ &\left(\frac{\partial T_L}{\partial z} + \frac{\partial T_U}{\partial z} - 12\frac{\partial T_m}{\partial z}\right) - \frac{\rho C_P h^2\left(u_L + u_U\right)}{2K}\frac{\partial T_m}{\partial x} - \frac{\rho C_P h^2\left(u_U - u_L\right)}{12K}\left(\frac{\partial T_U}{\partial x} - \frac{\partial T_L}{\partial x}\right) \\ &+ \frac{h^4}{12K\mu}\left[\left(\frac{\partial P}{\partial x}\right)^2 + \left(\frac{\partial P}{\partial z}\right)^2\right] + \frac{\mu\left(u_U - u_L\right)^2}{K} = 0 \end{aligned} \tag{12}$$

3.4. Heat conduction equation

The temperature in bush is determined by using the Laplace equation within the bearing material as given below Hori [2006]:

$$\frac{\partial^2 T_b}{\partial x^2} + \frac{\partial^2 T_b}{\partial y^2} + \frac{\partial^2 T_b}{\partial z^2} = 0 \tag{13}$$

In this equation, r stands for bush radius, and T_b stands for bush temperature. The equation (13) is then set into finite differences by using central difference technique. The final form is reproduced here.

$$\begin{aligned} T_b(i,j,k) &= E1T_b(i+1,j,k) + E1T_b(i-1,j,k) + E2T_b(i,j+1,k) + E2T_b(i,j-1,k) + \\ &E3T_b(i,j,k+1) + E3T_b(i,j,k-1) \end{aligned} \tag{14}$$

Where, $E = \left[\frac{2}{r^2 d\theta^2} + \frac{2}{dy^2} + \frac{2}{dz^2}\right]$; $F11 = \frac{1}{r^2 d\theta^2}$; $F22 = \frac{1}{dz^2}$; $F33 = \frac{1}{dy^2}$;

$$E1 = F11 / E; E2 = F22 / E; E3 = F33 / E$$

3.5. Computation procedure

Coupled numerical solutions of Reynolds, energy and heat conduction equations are obtained for offset-halves and elliptical journal bearings. The temperature of upper and lower bounding surfaces have been assumed constant throughout and set equal to oil inlet temperature for first iteration. For subsequent iterations the temperatures at oil bush interface are computed using heat conduction equations and appropriate boundary conditions. The numerical procedure adopted for obtaining the thermohydrodynamic solution is discussed below.

a. Reynolds Equation

A suitable initial value of attitude angle is assumed and film thickness equations (1-4) are solved. Then equation (6) has been used to find isothermal pressures. The initial viscosity values are assumed to be equal to the inlet oil viscosity.

b. Energy equation

The solution for the determination of temperature begins with the known pressure distributions obtained by solution of Reynolds equation. Viscosity variation in the fluid film domain corresponding to computed temperatures and pressures is calculated using equation (8). With new value of viscosity, equation (7) has been solved for thermal pressure. These values of pressure and viscosity, are used to further solve energy equation (12). Mean temperatures obtained by solving equation (12) are substituted in equation (11) to find the temperature in the oil film. Now, this temperature is used to solve the equation (13) to obtain the temperature variation in the bush. The computation is continued till converged solutions for thermal pressure loop and temperature loop have been arrived. The load carrying capacity is obtained by applying the Simpson rule to the pressure distribution. In computation, wherever reverse flow arises in domain, upwind differencing has been resorted to. Power losses have been evaluated by the determination of shear forces, and then employing the Simpson rule.

The boundary conditions used in the solution of governing equations are:

$P = 0$	at	$x = 0$	and	$x = l$
$u = u_L$	at	$y = 0$	and	$0 \le x \le l$
$u = 0$	at	$y = h$	and	$0 \le x \le l$
$T = T_0$	at	$x = 0$	and	$0 \le x \le h$
$T = T_L$	at	$y = 0$	and	$0 \le x \le l$
$T = T_U$	at	$y = h$	and	$0 \le x \le l$

$$T(0,y)=T_0\,;\; T(x,0)=T_0\,;\; k_{oil}\left(\frac{\partial T}{\partial y}\right)_{\substack{upper\\ bounding\\ surface}}=k_s\left(\frac{\partial T_s}{\partial y_s}\right)_{y_s=0}\;;$$

$$-k_s\left(\frac{\partial T_s}{\partial y_s}\right)_{y_s=t}=h_c\left(T_s\left(x_s,t\right)-T_a\right);\qquad -k_s\left(\frac{\partial T_s}{\partial x_s}\right)_{x_s=0}=h_c\left(T_s\left(0,y_s\right)-T_a\right)\;;$$

$$-k_s\left(\frac{\partial T_s}{\partial x_s}\right)_{x_s=l}=h_c\left(T_s\left(l,y_s\right)-T_a\right)\;;\qquad k_s\left(\frac{\partial T_s}{\partial z_s}\right)_{z_s=0}=h_c\left(T_s\left(x_s,y_s,0\right)-T_a\right);$$

$$-k_s\left(\frac{\partial T_s}{\partial z_s}\right)_{z_s=b}=h_c\left(T_s\left(x_s,y_s,b\right)-T_a\right)$$

Where, K_s denotes thermal conductivity of bearing, h_c denotes convection heat transfer coefficient of bush, l denotes length of the bearing, s denotes bearing surface, t denotes thickness of bearing, b denotes width of bearing, and T_a ambient temperature.

The solution of governing equations has been achieved by satisfying the convergence criterion given below:

For pressure:

$$\frac{\left|\left(\sum P_{i,j}\right)_{n-1}-\left(\sum P_{i,j}\right)_n\right|}{\left|\left(\sum P_{i,j}\right)_n\right|}\leq 0.0001 \tag{15}$$

For temperature:

$$\frac{\left|\left(\sum T_{i,j}\right)_{n-1}-\left(\sum T_{i,j}\right)_n\right|}{\left|\left(\sum T_{i,j}\right)_n\right|}\leq 0.0001 \tag{16}$$

Where, n represents number of iterations.

Here, the authors have made an attempt to present some performance parameters of offset-halves and elliptical journal bearing which have been evaluated using computer program developed by them based on method discussed in the previous articles of the chapter. Input parameters of offset-halves and elliptical journal bearing, and the properties of the grade oils and material used to manufacture the bearings are given in Tables 1, 2, and 3. The study has been carried at oil inlet temperature of $33\,^0C$ (which has been used as reference inlet temperature of the oil by most of the researchers) for the eccentricity ratios equal to 0.6 and speeds ranging from 3000 rpm -4000 rpm .

Parameter	Dimension	Tolerance	Roughness
Outer diameter of the bearing, OD	85mm	±0.2mm	10μm
Inner diameter of the bearing, ID	65mm	±0.2mm	10μm
Length, L	65mm	±0.2mm	10μm
Radial Clearance, C	500 μm	±50μm	
Minimum Clearance, C_m	200 μm	±50μm	
Oil hole	6.35mm	±0.15mm	
Relative sensor position	45º	±1º	

Table 1. Input parameters used to study the performance characteristics of offset-halves journal bearing [Chauhan: 2011]

Parameter	Dimension	Tolerance	Roughness
Outer diameter of the bearing, OD	85mm	±0.2mm	10μm
Maximum inner diameter of the bearing, D_{Imax}	65.4mm	±0.2mm	10μm
Minimum inner diameter of the bearing, D_{Imin}	65.2mm	±0.2mm	10μm
Length, L	65mm	±0.2mm	10μm
Radial Clearance, C	300 μm	±50μm	
Minimum Clearance, C_m	200 μm	±50μm	
Oil hole	6.35mm	±0.15mm	
Relative sensor position	45º	±1º	

Table 2. Input parameters used to study the performance characteristics of elliptical journal bearing [Chauhan: 2011]

	Oil 1 (Hydrol 68)	Oil 2 (Mak 2T)	Oil 3 (Mak Multigrade)
Viscosity, μ (at T_o=33 ^0C)	0.075 Pas	0.065 Pas	0.200 Pas
Viscosity, μ (at T_o=100 ^0C)	0.00771 Pas	0.004861 Pas	0.01239 Pas
Density, ρ	880 Kg/m^3	868 Kg/m^3	885 Kg/m^3
Thermal conductivity**, K_{oil}	0.126 $W/m\ ^0C$	0.126 $W/m\ ^0C$	0.126 $W/m\ ^0C$
Viscosity index*	98	135	110
Flash point*, ^0C	230	94	200
Pour point*, ^0C	-9	-24	-21
Barus viscosity-pressure index**, α	2.3e-8 Pa^{-1}		
Temperature viscosity- coefficient**, γ	0.034 K^{-1}		
Thermal conductivity of bush**, K_{bush}	0.22 $W/m\ Deg.C$		
Coefficient of thermal expansion of bush, h_{bush}	75e-6 K^{-1}		

Table 3. Properties of the bush material (Methyl Methacrylite) and grade oils under study [Chauhan et al. 2011]

Figures 5, and 6, show the variation of oil film temperature in the central plane of the bearing for eccentricity ratio, $\varepsilon = 0.6$ at oil inlet temperature of $33\,^0C$ for all the three grade oils under study at speeds 4000 rpm for offset-halves and elliptical journal bearing respectively. It has been observed that oil film temperature rise is very high in lower lobe in comparison to oil film temperature rise in upper lobe for offset-halves, and the oil film temperature rise is though high in lower lobe but it is comparable with the rise in upper lobe of the elliptical journal bearing. A high temperature rise in Oil 3 compared to the other grade oils has been observed which may be because of its high viscosity value. Similarly, figures 7 and 8, show the variation of thermal pressure in the central plane of the bearing for eccentricity ratio, $\varepsilon = 0.6$ at oil inlet temperature of $33\,^0C$ for all the three grade oils under study at speeds 4000 rpm for offset-halves and elliptical journal bearing respectively. It has been observed that thermal pressure rise is very high in lower lobe in comparison to thermal pressure rise in upper lobe for both offset-halves and elliptical journal bearing.

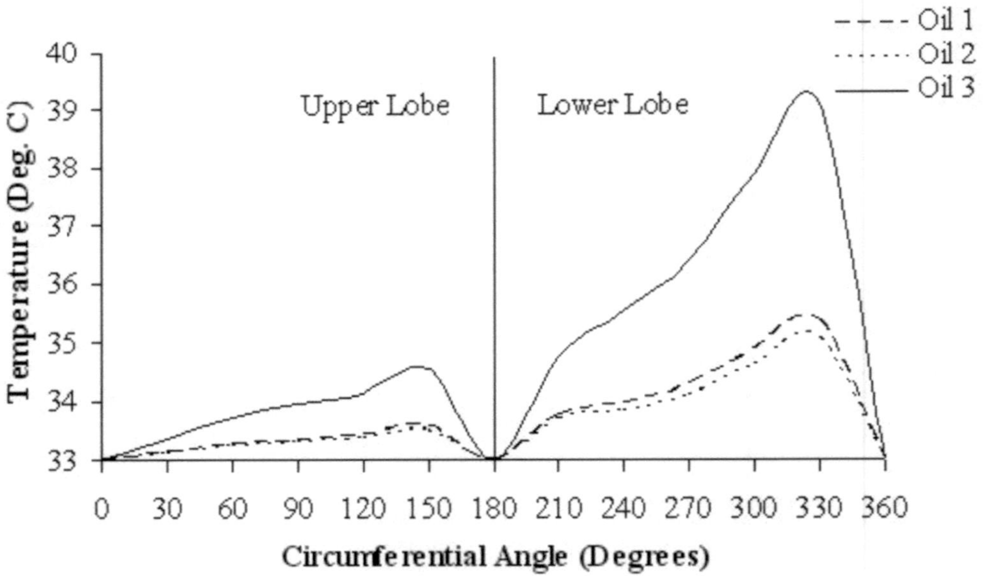

Figure 5. Variation of oil film temperatures in the central plane of the offset-halves bearing for different grade oils at 4000 rpm, oil inlet temperature=$33\,^0C$ and eccentricity ratio=0.6

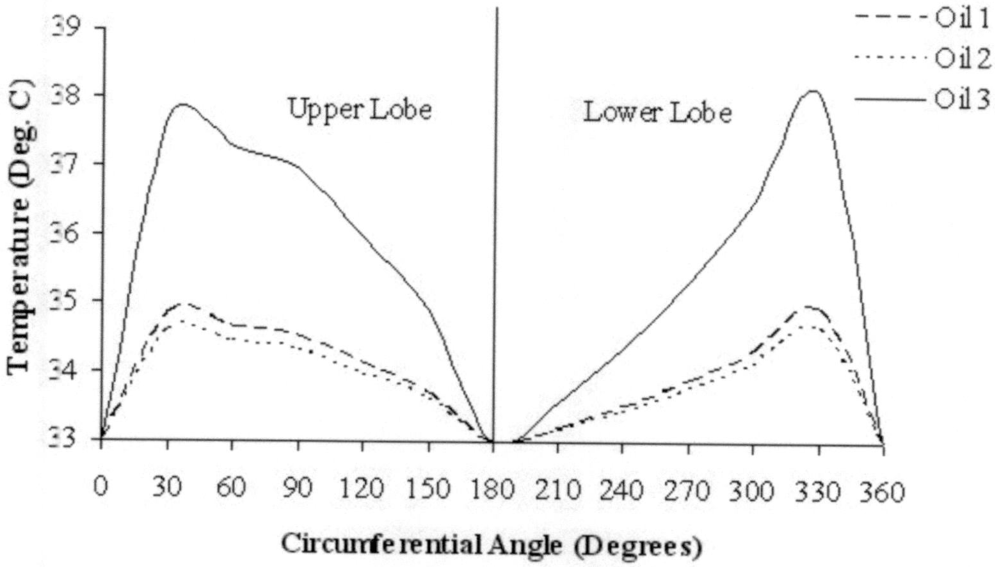

Figure 6. Variation of oil film temperatures in the central plane of the elliptical bearing for different grade oils at 4000 rpm, oil inlet temperature=33 ^{0}C and eccentricity ratio=0.6

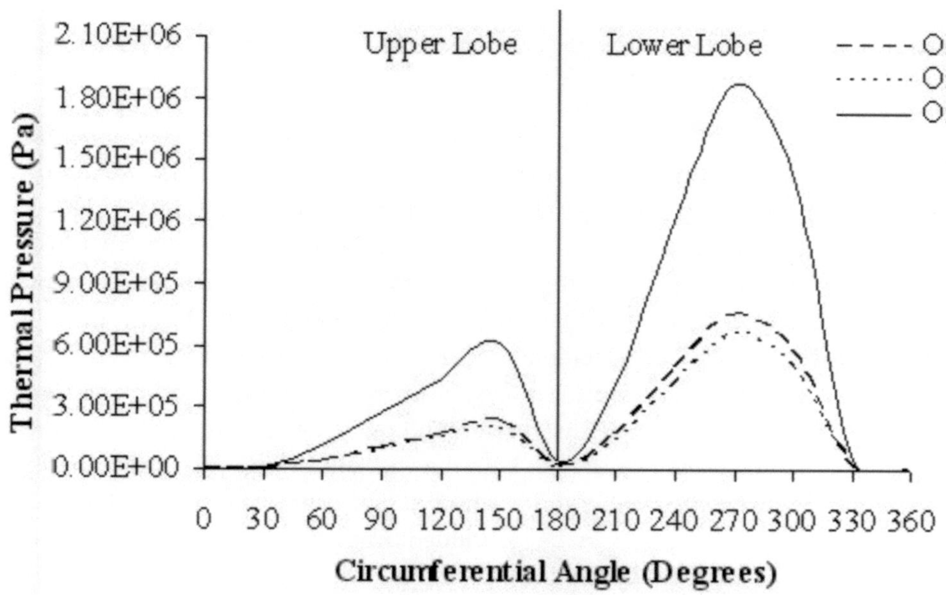

Figure 7. Variation of thermal pressure in the central plane of the offset-halves bearing for different grade oils at 4000 rpm, oil inlet temperature=33 ^{0}C and eccentricity ratio=0.6

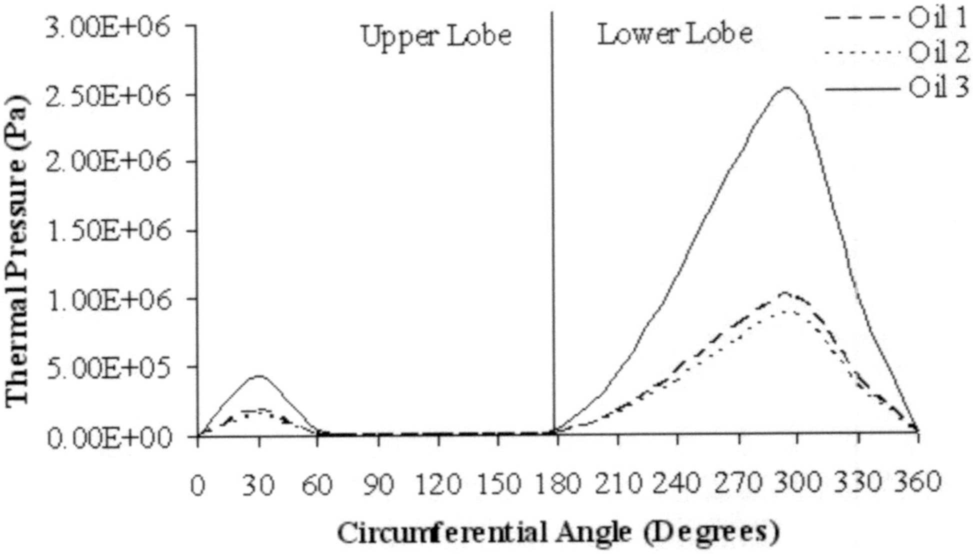

Figure 8. Variation of thermal pressure in the central plane of the elliptical bearing for different grade oils at 4000 rpm , oil inlet temperature=33 ^0C and eccentricity ratio=0.6

The comparative study of performance characteristics like oil film temperature, thermal pressure, load capacity, and power loss for the bearing configurations under study has also been presented in figures 9, 10, 11, and 12. Fig. 9 shows, that the oil film temperature rise is on much higher side for upper lobe of elliptical journal bearing when compared to the oil film temperature rise of same lobe of the offset-halves journal bearing, whereas the oil film temperature in lower lobe of offset-halves journal bearing is little higher the oil film temperature rise in lower lobe of elliptical journal bearing. Hence, the overall oil film temperature rise and thermal pressure has been observed high for elliptical journal bearing. The load capacity and power losses have been found on higher side for Oil 3 and on lower side for Oil 1. The trend remains same for both the journal bearings. The load capacity and power loss has also been observed high for elliptical journal bearing when compared with offset-halves journal bearing. It can be concluded from the above discussion that the lubricating oil with higher viscosity value results in high oil film temperature rise, high thermal pressure, high load capacity and also high power loss value, whereas the lubricating oil with low viscous value results in low oil film temperature rise, low thermal pressure, little low load capacity and power loss value.

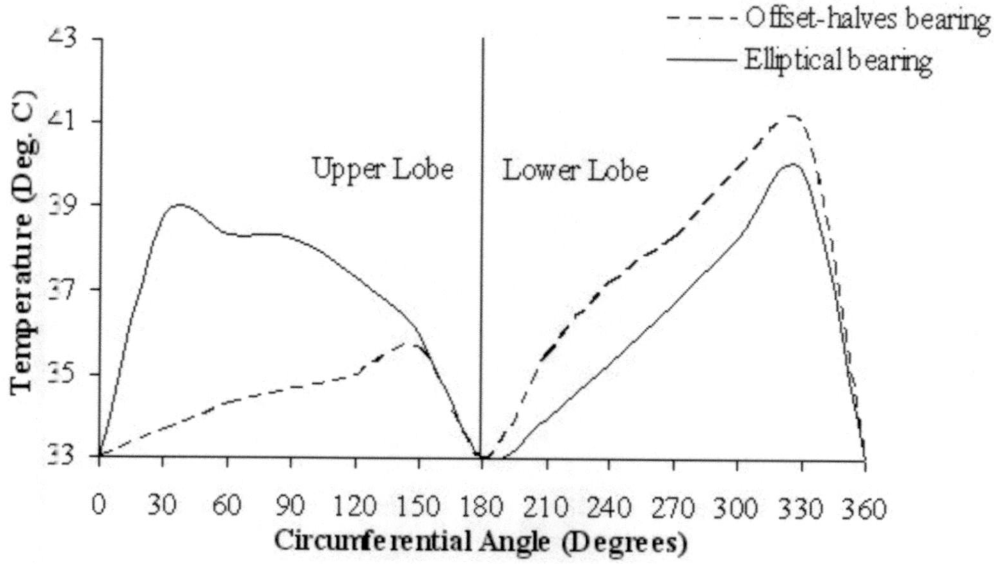

Figure 9 Variation of oil film temperature in the central plane with circumferential angle for Oil 1 at eccentricity ratio=0.6 and speed=3000 *rpm* for offset-halves and elliptical profile bearings (d= 0.1 *m*, l/d=1, $C = 200\mu m$, $C_m = 120\mu m$, T_o=33 ^{0}C)

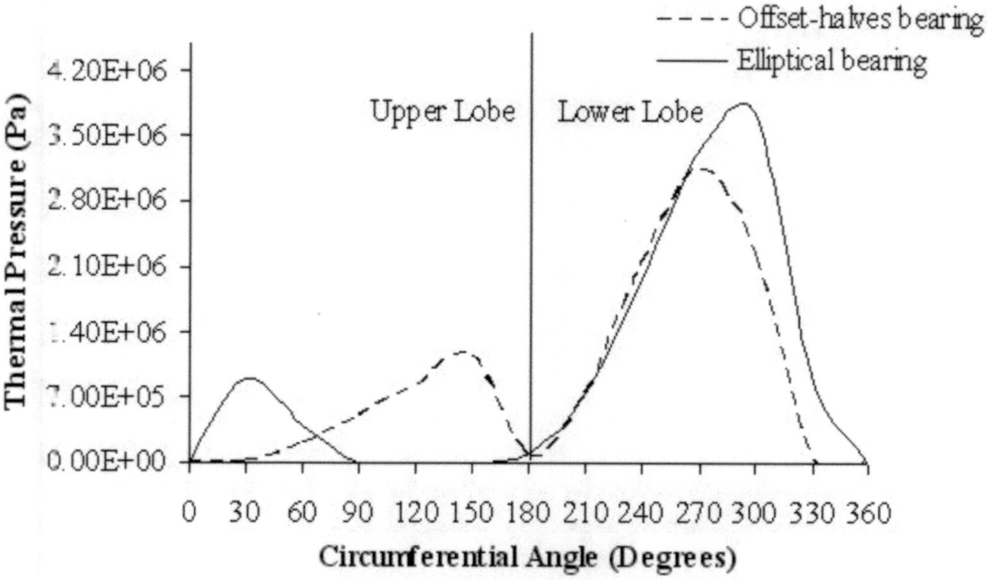

Figure 10. Variation of thermal pressure in the central plane with circumferential angle for Oil 1 at eccentricity ratio=0.6 and speed=3000 *rpm* for offset-halves and elliptical profile bearings (d= 0.1 *m*, l/d=1, $C = 200\mu m$, $C_m = 120\mu m$, T_o=33 ^{0}C)

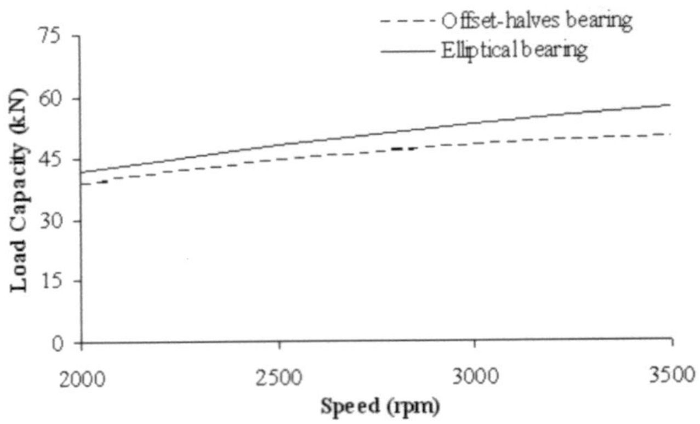

Figure 11. Variation of load capacity with speed for Oil 3 at eccentricity ratio=0.7 and speed=3000 rpm for offset-halves and elliptical profile bearings (d= 0.1 m , l/d=1, $C = 200\mu m$, $C_m = 120\mu m$, T_o =33 ^{0}C)

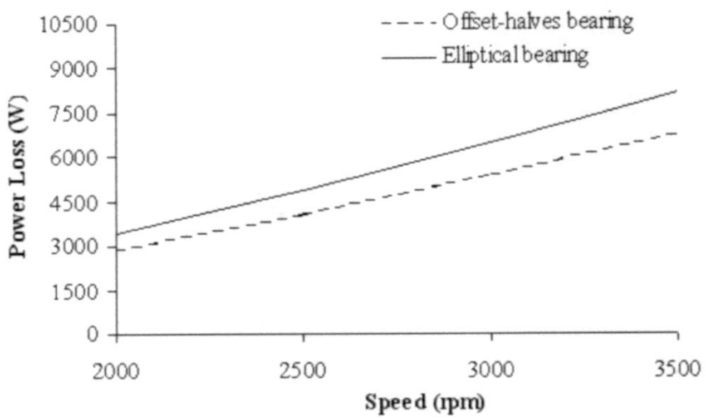

Figure 12. Variation of power loss with speed for Oil 3 at eccentricity ratio=0.7 and speed=3000 rpm for offset-halves and elliptical profile bearings (d= 0.1 m , l/d=1, $C = 200\mu m$, $C_m = 120\mu m$, T_o =33 ^{0}C)

4. Conclusion

This chapter deals with the thermal studies related to offset-halves and elliptical journal bearings. For each bearing configuration, the use of grade Oil 2 gives, the minimum temperature rise, and power losses which implies that the oil with low viscosity should be preferred as compared to high viscosity oil from thermal point of view and temperature rise is low for offset-halves bearing. An attempt has been made to compare the performance of two bearing configurations namely: offset-halves and elliptical journal bearings of same geometrical size by using three common commercial grade oils under similar operating conditions. Further, load carrying capacity of elliptical journal bearing has been found high in comparison to offset-halves journal bearing for same operating conditions. Among the two bearing configurations power losses are found to be minimum in case of offset-halves journal bearing. The chapter presents different performance parameters like oil film temperature, thermal pressures, load carrying capacity, and power losses which will help the designer to design this type of non-circular journal bearings as well as analyze their performance. The presently available design data/handbooks do not provide any direct analytical methods for the design and analysis of these non-circular journal bearings. However, the present methodology to a large extent has succeeded in standardizing the equations for design and analysis of these bearings.

Author details

Amit Chauhan
University Institute of Engineering and Technology, Sector-25, Panjab University, Chandigarh, India

Rakesh Sehgal
Department of Mechanical Engineering, NIT Hamirpur (HP), India

5. References

[1] Banwait SS., and Chandrawat HN., "Effect of misalignment on thermohydrodynamic analysis of elliptical journal bearings", IE (I) J 2000; 81: 93-101.
[2] Basri S., and Gethin DT., "A comparative study of the thermal behaviour of profile bore bearings", Tribology International, 1990; 23: 265-276.
[3] Booker JF., And Govindachar S., "Stability of offset journal bearing systems", Proc. of IMechE 1984; C283/84: 269-275.
[4] Chandra M., Malik M., and Sinhasan R., "Comparative study of four gas-lubricated non-circular journal bearing configurations", Tribology International, 1983; 16: 103-108.
[5] Chauhan A., "Experimental and theoretical investigations of the thermal behaviour of some non-circular journal bearing profiles", Ph. D Thesis, NIT Hamirpur, 2011.
[6] Chauhan A., and Sehgal R., "An experimentation investigation of the variation of oil temperatures in offset-halves journal bearing profile using different oils", Indian Journal of Tribology, 2008; 2: 27-41.
[7] Chauhan A., Sehgal R., and Sharma RK., "Thermohydrodynamic analysis of elliptical journal bearing with different grade oils", Tribology International, 2010; 43: 1970-1977.
[8] Chauhan A., Sehgal R. and Sharma RK., "Investigations on the Thermal Effects in Non-Circular Journal Bearings", Tribology International, 2011; 44; 1765-1773.
[9] Crosby WA., "Thermal considerations in the solution of finite journal bearings", Wear 1980; 64: 15-32.
[10] Fuller DD., "Theory and practice of lubrication for engineers", John Wiley and Sons, New York, 1956.
[11] Hori Y., "Hydrodynamic lubrication", Springer-Verlag Tokyo, 2006.
[12] Hussain A., Mistry K., Biswas S., and Athre K., "Thermal analysis of Non-circular bearings", Trans ASME 1996; 118: 246-254.
[13] Ma MT. and Taylor M., "An experimental investigation of thermal effects in circular and elliptical plain journal bearings. Tribology International, 1996; 29(1): 19-26.

[14] Mehta NP., and Singh A., "Stability analysis of Finite offset-halves pressure dam bearing", Trans ASME 1986; 108: 270-274.

[15] Mishra PC., Pandey RK., and Athre K., "Temperature profile of an elliptic bore journal bearing", Tribology International, 2007; 40(3): 453-458.

[16] Petrov NP., "Friction in machines and the effect of lubrication", Inzh. Zh. St. Peterburgo, 1883; 1: 71-140.

[17] Pinkus O., and Lynn, Mass., "Analysis of elliptical bearings", Trans ASME 1956; 55-LUB-22: 965-973.

[18] Rahmatabadi AD., Nekoeimehr M., and Rashidi R., "Micropolar lubricant effect on the performance of non-circular lobed bearings", Tribology International, 2010; 43: 404-413.

[19] Read LJ., and Flack RD., "Temperature, pressure and film thickness measurements for an offset half bearing", Wear 1987; 117(2): 197-210.

[20] Sehgal R., Swamy KNS., Athre K., and Biswas S., "A comparative study of the thermal behaviour of circular and non-circular journal bearings", Lub Sci 2000; 12(4): 329-344.

[21] Sharma RK., and Pandey RK., "An investigation into the validity of the temperature profile approximations across the film thickness in THD analysis of infinitely wide slider bearing", Tribol Onl 2006; 1: 19-24.

[22] Sharma RK., and Pandey RK., "Effects of the temperature profile approximations across the film thickness in Thermohydrodynamic analysis of Lubricating films", Ind J Tribol 2007; 2(1): 27-37.

[23] Singh A., and Gupta BK., "Stability limits of elliptical journal bearings supporting flexible rotors", Wear 1982; 77(2): 159-170.

[24] Singh A., and Gupta BK., "Stability analysis of orthogonally displaced bearings", Wear 1984; 97: 83-92.

[25] Singh DS., and Majumdar BC., "Computer-added design of hydrodynamic journal bearings considering thermal effects", Proc Inst Mech Eng, Part J: J Eng Tribol 2005; 219: 133-143.

[26] Singh DV., Sinhasan R., and Prabhakaran Nair K., "Elastothermohydrodynamic effects in Elliptical bearings", Tribology International, 1989; 22(1): 43-49.

[27] Singh DV., Sinhasan R., and Kumar A., "A variational solution of two lobe bearings", Mechanism and Machine Theory, 1977; 12: 323-330.

[28] Stachowiak GW., and Batchelor AW., "Engineering tribology", Elsevier Science Publishers B. V., Netherlands, 1993: 123-131.

[29] Tayal SP., Sinhasan R., and Singh DV., "Finite element analysis of elliptical bearings lubricated by a non-Newtonian fluid", Wear 1982; 80: 71-81.

[30] Tower B., "First report on friction experiments", Proceedings of IMechE, 1883: 632-659.

[31] Wilcock DF., "Orthogonally displaced bearings-I", ASLE Transactions, 1961; 4: 117-123.

[W1] www.roymech.co.uk/images3/lub_2.gif

[W2] www.reliabilitydirect.com/appnotes/jb.html

Section 2

Rolling Element Bearings

Chapter 2

Performance Evaluation of Rolling Element Bearings Based on Tribological Behaviour

Jerzy Nachimowicz and Marek Jałbrzykowski

Additional information is available at the end of the chapter

http://dx.doi.org/10.5772/51144

1. Introduction

It is a well-known fact that the appropriate test method could be one of the crucial conditions of the correct evaluation of a simulated process. One might presume that in some cases in which the subject matter is well-researched, only widely recognized methods are applied. However, in less familiar cases, or where one needs to evaluate some unusual values, the question of the appropriate methodology becomes far more significant. There are a number of papers in literature dealing with the analysis of the same phenomenon or process where tests were based on completely different methodological approach. As a result most papers quoted different results which caused serious controversy. In addition, it appears that even when the research was conducted according to the same procedure, but at different research units the results were still divergent.

As far as ball bearings are concerned there are numerous research methods to test these elements. For instance, using a four-ball pitting apparatus (T-03), a pitting ball-bearing tester (T-06), Amsler device or a ball-bearing testing platform. Most standards describe research methodology with the main focus on using particular testing devices, disregarding to a certain extent the very essence and purpose of the analysed object. The official standard for ball bearings in Poland is PN- 89/M-86410, however, it lacks a description of research methodology related to the external function of a ball bearing in „true-to-life" conditions. The description contains regulations concerning controlling the particular components of a ball bearing e.g. radial and axial run-out deviation, radial and axial clearance without paying any special attention to the working conditions of a ball bearing. Still, the standard allows applying other research methods which require prior acceptance of both the manufacturer and recipient of ball bearings.

In this context, the authors of the present section conducted tests on two-set ball bearings according to two different methods: their own methodology and the methods recommended by the manufacturer (FŁT PLC in Kraśnik - Poland). Should add that all of the test, later in this chapter bearings are factory items and their designations are in accordance with PN, PN-EN and ISO standards. All tests were performed for three replicates at each point.

1.1. The metods of investigations

Tribological tests were conducted with a widely-used friction tester SMT-1. Fig. 1 shows the apparatus and the structure of the friction node.

Laboratory tests were conducted in two stages. In the first stage tribological tests were conducted according to the authors' own methododology. Next the tests were conducted according to the method applied in FŁT in Kraśnik (Poland). These tests were meant to compare the two methods and evaluate their influence on test results. It should be noted that the load is strictly radial, without a longitudinal component which has been verified experimentally, and this is due to the construction SMT1.

1.1.1. Authors' own methodology (MB –1)

In this section the following test parameters were assumed: constant normal pressure N = 350 [kG] ≈ 3433 [N], changeable rotational speed of the bearing : n_1= 300 [rpm], n_2= 600 [rpm], n_3= 1200 [rpm]. Because of changeable rotational speed it was assumed that the number of bearing's cycles will serve as the unit of its operation. Since the tests were treated as pilot ones, at this stage tests were conducted at 20,000 cycles of bearing's operation. It should be noted that before each measurement the bearing was cooled in a stream of air for 4 hours in normal conditions. All the measurements were recorded using a computer system.

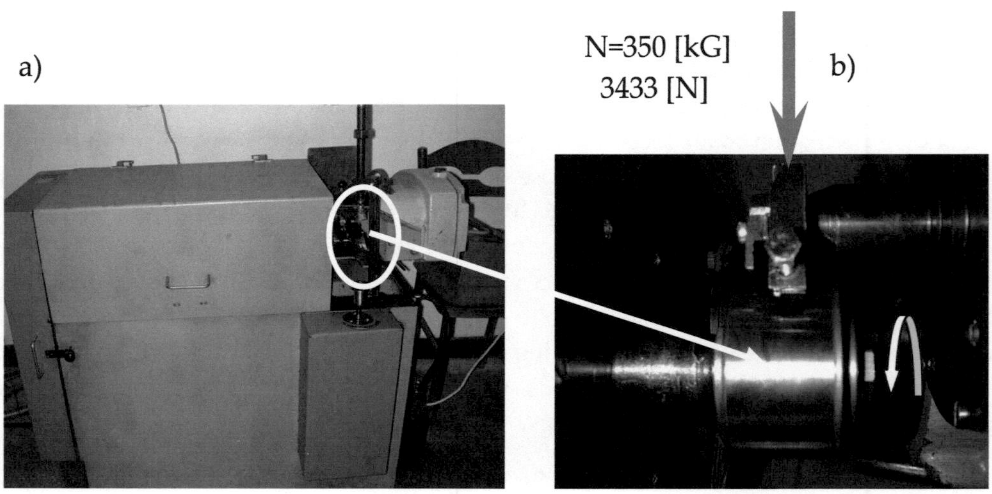

Figure 1. SMT-1 friction tester a) general view, b) friction node

1.1.2. Tests recommended by FŁT in Kraśnik (MB-2)

Before the tests were started, in the initial stage, normal force of N= 350 [kG] ≈ 3433 [N] was applied to the bearing. Next the apparatus was switched on for t=30 [min] at the speed of n= 1000 [rpm]. After that the rotational speed was increased to n= 1400 [rpm] and as soon as the system became stable the moment of friction was measured (M_t). Then, the rotational speed was decreased by 200 [rpm] in each step, and on each occasion after the friction node became stable the measurements M_t were taken.

1.2. Test results and discussion

Due to the extensive research material, the present section does not contain the results of initial tests obtained with MB-1 and MB-2 methods, it only includes their description.

As far as MB-1 method is concerned, it should be noted that its results indicate diversification of tribological characteristics of the examined elements in relation to rotational speed. There is also a certain analogy in the course of characteristics for the two groups of the evaluated bearings. In the initial stage of the process the maximum value of friction moment was obtained (for all rotational speeds). In the subsequent stages it decreases and, in the final stages, the resistance of motion becomes stable. One should note that at the beginning of the process the highest value of the moment of friction was obtained at the speed of n = 1200 [rpm] while the lowest at n = 300 [rpm]. In the final stage of the process the trend is reversed.

In the case of MB-2 method general diversification of the moment of friction was also noticed. It is logical considering different conditions of the external function. However the lowest value of the moment of friction was recorded for the lowest rotational speed n = 200 [rpm], while the highest for n = 1000 [rpm]. Further measurements, at higher rotational speeds, indicate the decrease of the value of the moment of friction. However, the most controversial conclusions are drawn from the comparative analysis of the two research methods.

Fig. 2 and 3 shows selected comparative graphs of the courses and values of the moment of friction obtained in tests according to MB-1 and MB-2 methods described above.

The graphs above provide data concerning completely divergent results in the conducted tests. The analysis of the data in fig. 2. (MB-1 method) proves that both at the speed of n = 300 [rpm], and especially n = 1200, throughout the whole of the research cycle, one notices distinctly higher values of the moment of friction for the bearings from the 134-781TNG-2RS group. At the same time there are lower values of resistance to motion for the bearings from the CBK 441TNG group. Since these bearings have different dimensions of working elements one could assume that different values of the moment of friction is a natural phenomenon. When tests were carried out according to the other method similar trends in friction characteristics were expected. However, when the results obtained with MB-2 were analysed, quite contrary to expectations, they were in direct opposition to those obtained with MB-1 method.

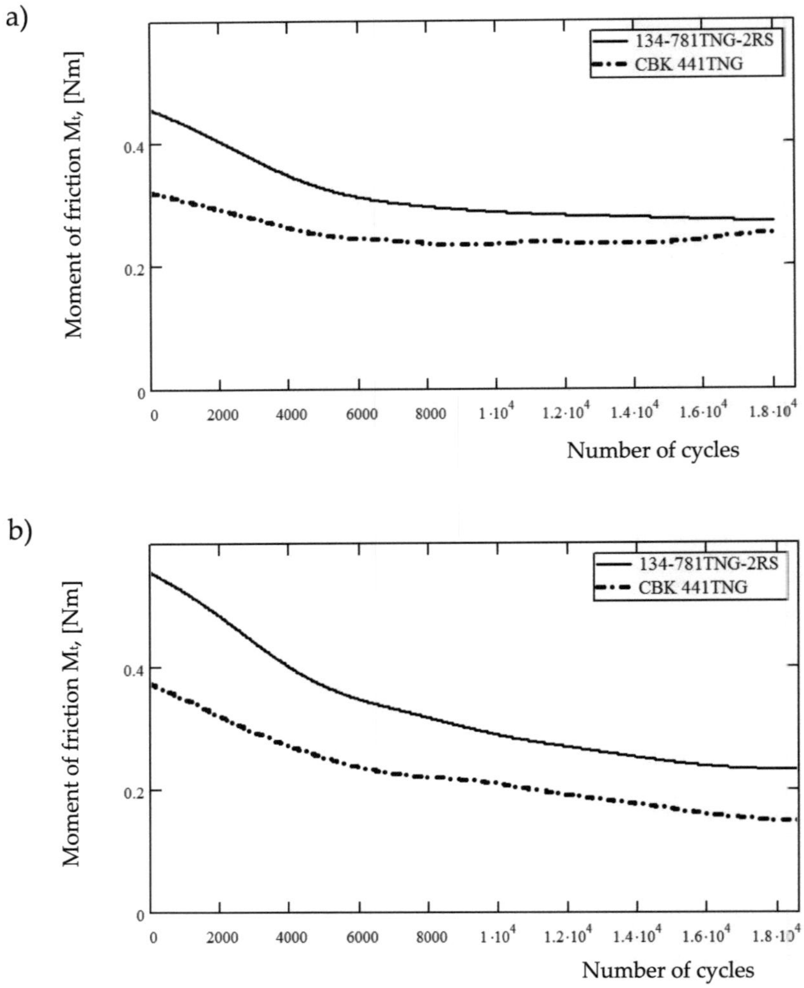

Figure 2. The course of the moment of friction in the function of time for two groups of bearings at different rotational speeds (MB1 method): a) 300 [rpm], b) 1200 [rpm]

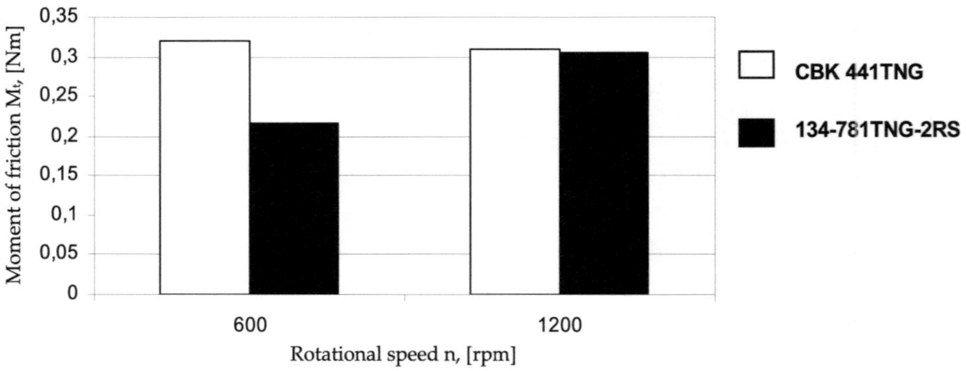

Figure 3. The values of the moment of friction in the function of rotational speed – MB-2 method

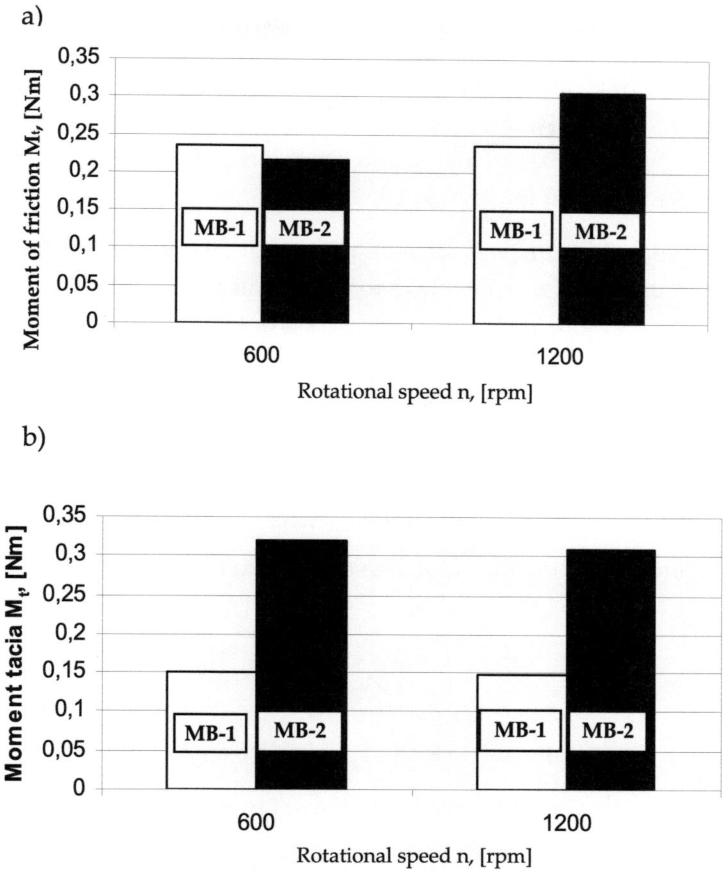

Figure 4 Comparison of two research methods MB-1 and MB-2 (bearings) a) 134-781TNG-2RS b) CBK 441TNG

This comparison is shown in fig.4. Fig. 4a refers to the test results for the bearings from the 134-781TNG-2RS group, fig. 4b is related to the bearings from the CBK 441TNG group. No doubt, the results for the two groups of bearings are completely different. It should be noted that, in the 134-781TNG-2RS group, the differences in the recorded values of the moment of friction range from several to twenty-odd per cent, while in the CBK 441TNG group the differences are even two-fold. Moreover, one should note that in the case of bearings from the 134-781TNG-2RS group at the speed of n = 600 [rpm] slightly higher values of the moment of friction were registered with MB-1 than with MB-2 method. In turn, at the speed of n = 1200 [rpm] the situation is reversed. That means that when tribological tests are carried out, especially on such elements, one should bear in mind the significance of the selected research method Choosing an arbitrary method may result in, as the evidence discussed in the present section suggests, obtaining entirely different or even unexpected test results. In consequence it may lead to wrong conclusions related to the tribological properties of such elements. At the same time one should remember it is often a complex task to select the right method that can reflect the real process of technical objects exploitation closely enough.

2. Selected energetic aspects of needle bearings work performance

Shafts on needle bearing can work at heavy load accurately lateral. Durability of such a bearing is guaranteed by suitable selection materials which are made (roughness, microhardness, etc.) and optimum lubrication conditions. The factors specified above have influence on reliable work and life of needle bearings.

Wear equation presented in energetic form describes real processes occurring under various external influences on materials which lead to their damage and create wear products. The initial form of energetic approach to wear processes is as follows:

$$E_{y\partial} = A_{wn} / V \tag{1}$$

where:

$E_{y\partial}$ - specific energy of wear material of volume V when wear products are created, A_{wn} - the work of external forces affecting the volume in time t or at N cycles of load.

It is assumed that:

$$A_{wn} = A_i N = A_i \omega t \tag{2}$$

where: A_i – external work under single load of volume V, ω - frequency of load.

Equations (1) and (2) result in wear equation in the following form:

$$V = \frac{A_i}{E_{y\partial}} \omega t \tag{3}$$

Equation (3) describes the wear when wear products are produced at a uniform rate i.e. linear wear.

For non-linear wear:

$$V = \frac{A_i}{E_{y\partial}} f(\omega t) \tag{4}$$

where: the function f (ω t) may be presented (apart from the linear one) as power, exponential, logarithmic and as other relations.

In dynamic character of load (impact, hydro-abrasive, cavitation wear) the equation (4) should consider the velocity of impact:

$$V = \frac{A_i \vartheta_i}{E_{y\partial} \vartheta_{kr}} f(\omega t) \tag{5}$$

where:

ϑ_i - the average value of velocity of impact load, ϑ_{kr} - critical velocity of impact which causes damage of wear material , $A_i\vartheta_i$ – the flux of energy which enters a material in the course of separate impact, $E_{y\partial}$ ϑ_{kr} – critical density of strain energy – critical density of strain power.

Assuming $A_i\vartheta_i$ = w and $E_{y\partial}$ ϑ_{kr} = W_{kr}^* , we obtain:

$$V = \frac{w_i}{W_{kr}^*} f(\omega t) \tag{6}$$

The numerator and denominator of this equation are reduced after the singularity of a given type of wear is taken into account. The flux of external energy w should be expressed by the friction force in external friction. Fig. 5 shows the model of distribution of the energy of plastic deformation.

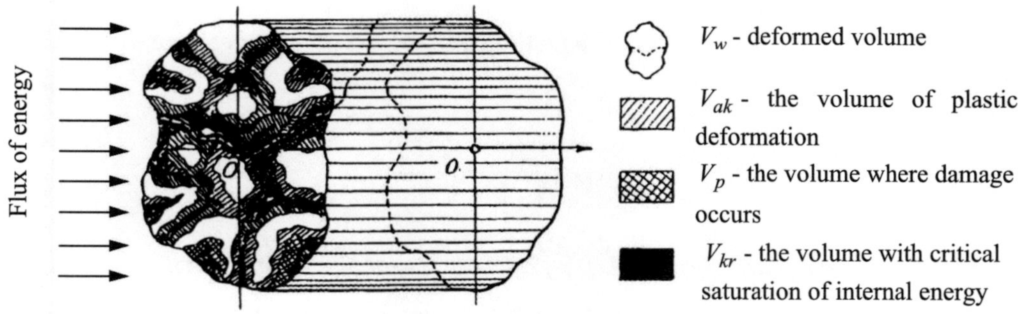

Figure 5. The model of the energy flux distribution in the deformed volume

The criterion of resistance to wear W_{kr}^* should consider the properties of materials (mechanical, physical, chemical and others), causing damage in given conditions of external influences. The value W_{kr}^* depends on the character of external load and assumes the averaged values in the worn out volume (* the symbol of averaging).

By presenting ϑ_{kr} as the sum of components of elastic and plastic deformation, ϑ_{kr}^{spr} ϑ_{kr}^{pl} respectively, we shall obtain:

$$W_{kr}^* = \frac{1}{3} E_{y\partial}^* \vartheta_{kr} = \frac{1}{3} E_{y\partial}^* \left(\vartheta_{kr}^{spr} + \vartheta_{kr}^{pl} \right) \tag{7}$$

The specific strain energy E_{ya}^*, irreversibly absorbed by the material at the instant of creating wear products, can be determined according to the durability or hardness graph as an area below the curve $\sigma(e_i)$ or H (e_i), restricted on the right side by the damage deformation e_p determined in given wear conditions with the method of micro-hardness directly on the worn out surface of a material.

The elastic component ϑ_{kr}^{spr} in the equation (7) is defined with the velocity of the elastic wave in deformation c_0 and elastic deformation e_{spr}:

$$\vartheta_{kr}^{spr} = c_0 e_{spr} = \sqrt{\frac{E}{\rho_M} e_{spr}^2} \tag{8}$$

where:

E – the module of elasticity, ρ_M - density of a material being used. Plastic component ϑ_{kr}^{pl} in (7) depends on the velocity of plastic deformation wave c_{pl} and total plastic deformation accumulated in the course of wear:

$$\vartheta_{kr}^{pl} = \int_{e_{spr}}^{e_p} c_{pl} de = \int_{e_{spr}}^{e_p} \left(\frac{d\sigma/de}{\rho_M} \right)^{0.5} de \tag{9}$$

where: $d\sigma/de$ - local slope of tangent to a curve of stretching determined with stress – strain coordinates (σ - ε).

By expressing the right-hand side of equations (8) and (9) as the energy of brittle E_{kr} and plastic E_{pl} damage we shall obtain:

$$\vartheta_{kr} = \sqrt{\frac{2E_{kr}}{\rho_M}} + \sqrt{\frac{2E_{pl}}{\rho_M}} \tag{10}$$

where: E_{kr} = $E_{kr} = \sigma_b^2 / 2E$, $E_{pl} = (\sigma_b - \sigma_T)^2 / 2D$, σ_b, σ_T - strength of a material and the limit of plasticity, D – module of consolidation (tg of inclination angle of tangent to a curve $\sigma(e)$ within the range of plastic deformations, with linear character of deformation consolidation of a material. The critical velocity of impact, connected with singularities of spreading in materials undergoing wear, elastic and plastic deformation waves, features resistance to material wear at dynamic, high-velocity stress and it can be expressed by durability characteristics of worn-out volumes, as shown in equations (8) – (10).

Test data analyses proves that in conditions of stretching at impact (impact on the front surface of a rod) there exists an involution dependence (between ϑ_{kr} i σ_b):

$$\vartheta_{kr} = const \sigma_b^n \tag{11}$$

where: n = 2,5 for steel with $\sigma_b \leq 500$ [MPa]

const – a constant determined empirically, considering the specific test conditions or the proportional coefficient between the discussed characteristics.

Taking into account (11) the criterion of resistance to wear for iron-based alloys with $\sigma_b \leq 500$ [MPa], it will prove that ϑ_{kr} and σ_b cubed are proportional:

$$W^*_{kr} \approx const\vartheta^3_{kr} \approx const\sigma^3_b \tag{12}$$

The particular dependencies (12) are of significant practical interest with a view of predicting the use of materials and coating. By substituting equation (10) for (7) we shall obtain:

$$W^*_{kr} = \frac{1}{3}E^*_{y\partial}\left(\sqrt{\frac{2E_{kr}}{\rho_M}} + \sqrt{\frac{2E_{pl}}{\rho_M}}\right) \tag{13}$$

In this equation, the total energy-consumption of a material E_{ya}^* and its components E_{kr} and E_{pl} can be determined according to a microhardness (endurance) graph. If the graphs of hardness or endurance cannot determined, then the evaluation of resistance of materials to wear at external impact is possible as a result of analogy to W_{kr}^* criterion.

Still, it is observed that when the stiff impact stress is reduced and when external influence resemble static conditions, W_{kr}^* criterion will assume very simple forms. For instance it can be expressed in reduction of indicator in the power at ϑ_{kr} and σ_b or hardness (12), [5, 6, 8].

2.1. The method of investigation

Fig. 6 shows the scheme of the tribotester for the estimation of needle bearing. In this case, used the bath lubrication. Bearing designation are given in Fig. 6.

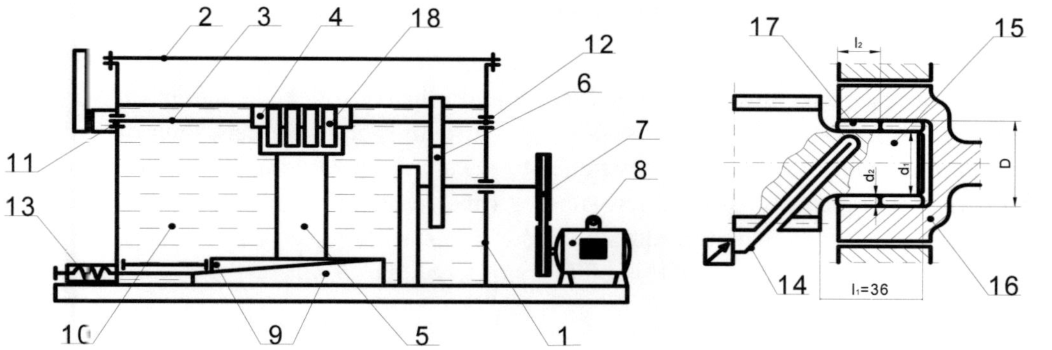

Figure 6. The scheme of the tribotester: 1- frame, 2- cover, 3- investigated shaft, 4- reducer sleeve, 5- cylinder, 6- toothed gear, 7- transmission belt, 8- electric motor SZJe 3,5 [kW], 9- wedges, 10- oil, 11- needle bearing K28x33x13, 12- needle bearing K40x45x17, 13- screw gear, 14- thermoelement NiCrNi, 15- shaft neck d= 28 [mm], 16- hub, 17- needle roller ϕ2,5 [mm], l2= 13 [mm], 18- bearing head Parameters of investigation:surface stresses, p = 950 [MPa], rotational speed, n = 1500 [rpm].

2.2. The results of the investigations

Fig. 7 shows the influence of energetic criterion on the wear of a shaft neck in the function of change of material and lubricating parameters.

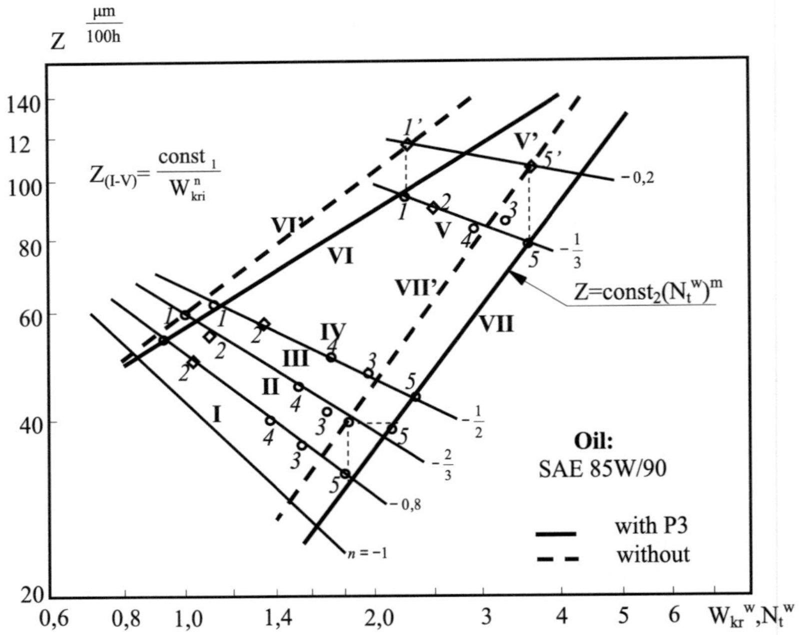

Figure 7. Wear intensity of a shaft neck – energetic criterion: N_t^w – relative density of friction power; W_{kr}^w – relative critical density of strain power ; a shaft neck: 1 – 16MnCr5, 2 – 20MoCrS4 + carburising; welding : 3-80G; 4-35X5M1,5; 5-50X10GC1,5; P3 – addition to Acorox 88 oil, needle bearing K 40 x 45x17

The analysis of test data shows that at general resistance to wear criterion for materials at external friction there occurs average critical density of friction power W_{kr}^* , in worn out volumes.

If chemical reactions, surface-active substance and other factors have significant influence on the processes of creating various secondary structures and products, then determining adequate reliability of criteria of resistance to wear proves to be a difficult problem, for instance at low-intensity oxidizing wear of a tribo-coupling. Resistance criteria in such conditions might be: dislocation density in the thin surface layer of a material, activating energy of chemical reactions and durability characteristics [5 ,9, 10].

The list of materials resistance to wear criteria – invariants of universal criterion W_{kr}^* (critical density of strain power) indicates huge variety and complexity of processes of materials surface damage.

It ought to be noted that the properties which characterize deformed areas of materials at the moment of their destruction (the initial stage of creating wear products) show reactions of these areas to external influence. In order to point those reactions in the desired direction, in this case – to ensure high resistance to abrasion (endurance) of its elements, it is necessary to make the right choice of known materials or produce new ones, having structure of the highest resistance to cracking resulting from the influence of external (exploitation) factors. This means that materials science involving problems of tribotechnology should be based on the analysis of microstructures.

3. The influence of geometrical parameters on the friction process in the needle bearing

The construction of the rolling bearing was initially based on the assumption that the friction loss during the bearing work is significantly smaller than during the sliding. However, during the work of the bearing in the operating conditions there exists both the bearing and the sliding friction. Different factors result in appearance of resistance to motion while the bearing is operating:

- Hysteresis of deformations;
- Interior friction in lubricant;
- Sliding and microsliding caused by deformations, the geometry of contact area and the movement caused by the gyroscope moment;
- Sliding between the bearing cage, the bearing elements and the bearing ring;
- Sliding in the bearing seal.

When bodies are deformed in the operating conditions, the phenomenon of pure bearing exists if the cooperating elements possess the same diameter, length, the properties of the material and parallel axes. Also, the roughness of the cooperating surfaces should be minimal. In such a case, when there is no lubricant, there appear only some losses caused by hysteresis of deformations [15]. This stems from the fact that the difference between the length of the contact arc and the corresponding arcs before deformations is identical for both bodies. As a result, there is no mutual sliding of the surfaces in the deformation area.

If the curvatures of the cooperating surfaces are different, in the elastic deflection the length of the contact arc for both bodies is identical, whereas before the deflection it was different. Consequently, the deformations in the contact area are accompanied by microsliding; if, however, the speed of both cooperating surfaces is identical, their mutual movement is called „bearing".

In a typical situation, in the process of bearing of two bodies with different peripheral speed there occurs the bearing with sliding; such a situation is the subject of examination here. The aim is to estimate of the extent to which the sliding friction matters in the overall balance of motion resistance.

3.1. The analysis of contact areas of the needle bearing elements

In the place of contact of two elastic bodies pressed against each other with some force, some contact stresses within a certain field of mutual contact occur. They reach significant values even in the situation when the pressing force is relatively small, which, as a consequence, may lead to exceeding the acceptable limit of the material effort. This is of paramount importance during the work of needle bearings which are under considerable load. Figure 8b shows a situation when the axis of the needle and the axis of the shaft neck are parallel. The stresses that occur (Fig. 8 a) are evenly spread along the length of the needle, and the area of contact between the two elements equals the field of the ellipse of the length which is the same as the length of the needle and the width 2b calculated by means of Hertz's solution:

$$b = \sqrt{P \frac{D_1 \cdot D_2}{D_1 + D_2} \left(\frac{1-v_1^2}{E_1} + \frac{1-v_2^2}{E_2} \right)} \tag{14}$$

where *P* is the strength per each unit of the rollers length; *D1, D2* are the diameters of the shaft neck and the needle respectively; *v1,v2* are Poisson's figures; *E1,E2* are Young's modulus. The indexes *1,2* refer to the roller *1* (the shaft neck) and *2* (the needle roller) respectively.

In the case when the axes of the touching rollers (the shaft neck and the needle) are not parallel (Fig. 8 c), the whole load and the stress connected with it concentrate in a relatively small point [13].

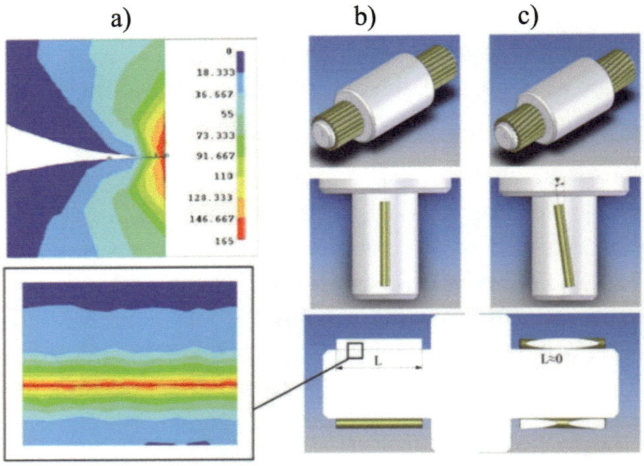

Figure 8. The pictorial diagram of the roller bearing: a) the spread of stresses for the connecting rollers with parallel axes, b) the axes of the parallel elements, c) the axes of the elements shifted by an angle

The area of the friction surface of the elements of bearing is influenced not only by the adequate mutual positioning of the cooperating elements, but also by the changeable relation of the diameters of the needle and the roller, changeable values of Young's modulus, Poisson's ratio of the materials used, and the change of load.

The basic theoretical perspective assumed while calculating contact stresses and the width of the contact area between co-working bodies was based on Hertz's theory, drawing on the following premises: the contacting elements are made from a homogenous, isotopic material; they are limited by the smooth surfaces with a regular curvature; and within the contact area some deformations occur [11, 12, 14].

The analysis of the change in the contact area of the elements of needle bearing presented below was prepared on the basis of all the premises of Hertz's theory concerning the work of two rollers (the shaft neck, the needle roller) with parallel axes, working under static load in dry environment. In the examination of the factors which have an impact on the measure of contact area of the elements cooperating in the form of 'the shaft neck – the needle roller' interaction is particularly important as the contact area influences the change in the moment of motion resistance in the process of bearing.

In the first test eight different needles of 1,4 [mm] up to 2,5 [mm] in diameter were juxtaposed with four rollers of 10, 20, 30, 40 [mm] in diameter. The contact area of the cooperating elements was calculated taking as an assumption the constant load of 200 [N/mm]. Different combinations of elements (needles and rollers) were tested. The bigger the diameter of the needle, the bigger the contact area becomes – this tendency can be measured by means of power equations (Fig. 9.).

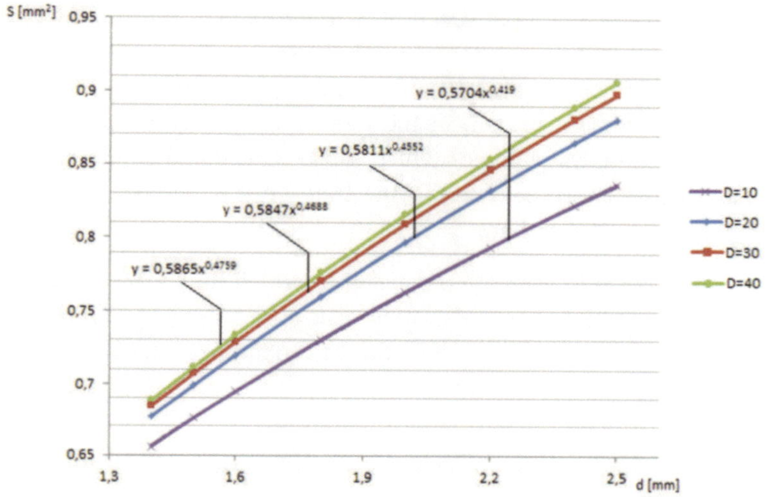

Figure 9. The influence of the needle roller diameter (d) and the shaft neck diameter (D) on the contact area (S) of the elements of bearing.

In the case demonstrated above, the biggest changes in the contact area can be observed while putting together the needles of bigger diameters with the given set of rollers, i.e. for the needle of 1,4 [mm] in diameter cooperating with the rollers of 10 and 40 [mm] in diameter the contact area increased by approximately 5%, while for the needle of 2,5 [mm] in diameter cooperating with the same rollers the contact area increased by approximately 9 [%].

3.2. The mathematical model of the friction process in the needle bearing

The state of the present studies, the information in scientific publications and the catalogues of leading producers of bearings show that the total moment of resistance in the work of a bearing can be understood as the sum of the elemental moments originating in the bearing friction, the sliding friction, the bearing seal friction and the friction occurring in oil environment.

However, in order to estimate the moment of resistance in the friction sliding we suggest the mathematical model in which the total moment of different types of resistance to motion comprises firstly the moment resulting from the bearing friction, calculated as the product of the load – N and the rolling friction ratio - f, and the moment originating from the sliding friction, taking into account load, the sliding friction - μ and the radius of the roller – r (2).

$$M_{Tc} = N \cdot f + N \cdot \mu \cdot r \qquad (15)$$

The recommended examination on a stand (the friction machine SMT-1) makes it possible to obtain the values of the moment of resistance created in the process of sliding friction from the total moment of resistance to motion in the bearing. The needle bearing with the outer ring is mounted directly on the shaft neck which is an interior bearing track for the needles. The outer ring is stationary, and the bearing is loaded with radius force. During the work of the roller turning with the rotational speed of 50-100 rotations per minute there exists only the resistance connected with bearing friction. The observations provided by a torque meter help estimate the moment of bearing friction and the value of the ratio *f*. Then, if the number of rotations of the roller is increased, in the total moment there will appear the already mentioned sliding friction, the value of which can be calculated as the difference between the total moment of resistance obtained in the trail work of the bearing and the value of bearing resistance established earlier.

Such an examination helps to estimate the extent to which the bearing friction and sliding friction matter in the overall balance of motion resistance. The next step in the examination may be to check the influence of load with constant rotational speeds, and the influence of the change in the lubricant environment.

3.3. The analysis of the friction process as the function of the initial slackness

The bearing slackness is a very important geometrical parameter which determines the intensity and the type of wear. In order to estimate the influence of slackness on the process of wear the examination was carried out for the nominal slackness of 0.065 [mm] and other slacknesses achieved as a sum of the nominal one and the value of reduction of the shaft neck diameter. The associations received in this way made it possible to examine the process of friction with slacknesses of 0.1, 0.2, 0.3, 0.4, 0.5 [mm] respectively. A hundred-hour examination was carried out (on the friction machine SMT-1) for each value of slackness in bearing; next, the micro-structure of the friction surface was examined under an optic microscope with magnification X300 (Fig. 10 b). On the basis of the results received it can established that the enlargement of slackness between the elements in the friction pair leads to a considerable enlargement of the intensity in the shaft neck wear, which, in turn, results in the worse durability of the connection.

Fig. 11. shows the kinetic relations between the wear of the shaft neck and different slacknesses in the function of work time. The lines 1 and 2, corresponding to the slacknesses 0,5 [mm] and 0,4 [mm] respectively, indicate the group with a relatively high rate of wear. A significantly smaller rate of wear is represented by the lines 4-7, corresponding to the slacknesses 0,065 [mm] – 0,3 [mm] respectively (Fig. 11). The dotted line 3, typified as the critical slackness h = 0,325 [mm] characterizes the shift from the less intensive, mechano-chemical one, to the more intensive one. The increase of the slackness from the critical h = 0,325 [mm] to 0,4 [mm] results in the intensity which is approximately twice bigger. With the slackness 0,5 [mm] the durability of the needle bearing decreases by 50 [%], if compared with the durability with the critical slackness. Fig. 10 shows the wear of the shaft neck understood as function of the initial slackness. Point A (Fig. 10 a) as the critical slackness with which the friction pair works in the conditions of normal wear was achieved as a result of crossing the tangents led to the curve of wear in areas I and III [13, 15].

a)

b)

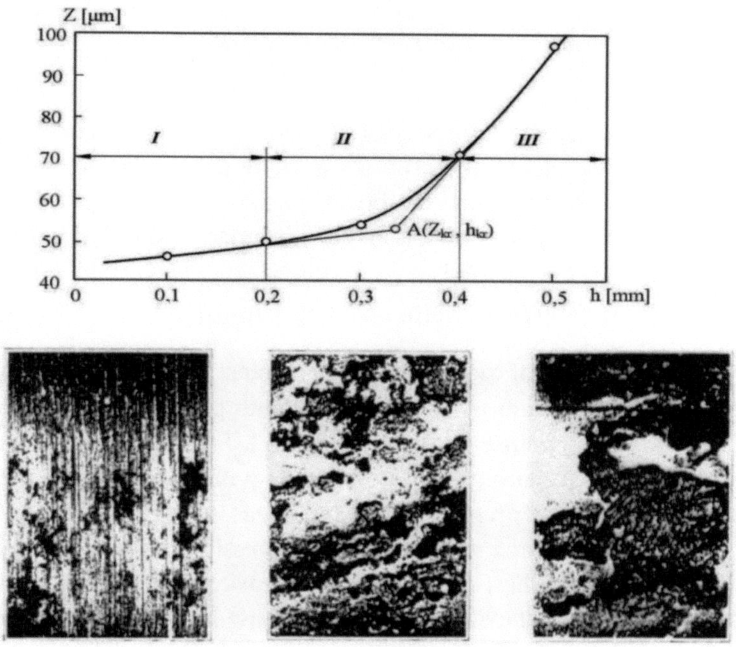

Figure 10. Demonstration of the shaft wear: a) the shaft wear in the function of initial slackness where: I is normal wear, II is mechano-chemical wear, III is pathological wear, b) the structure of the worn surface of the shaft neck x 300

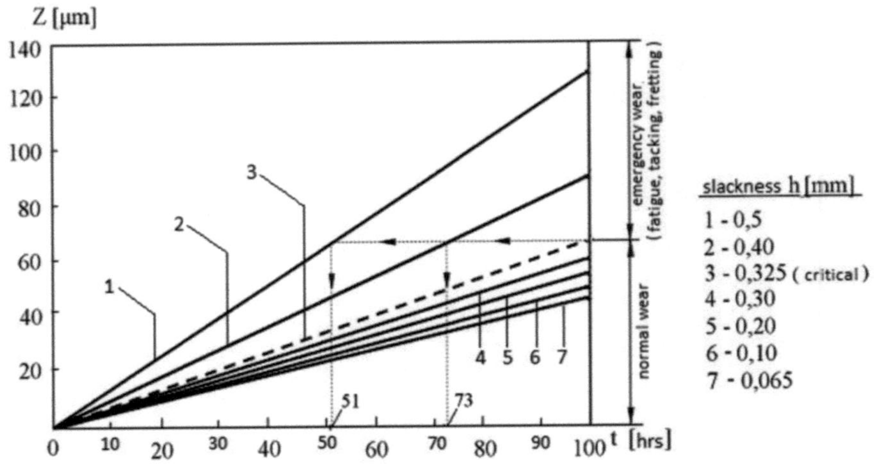

Figure 11. The kinetic relations in the wear of the needle bearing elements (the operating conditions, different slacknesses in association)

In real conditions, when the surfaces of the bodies in direct contact are uneven, and anisotropy of top layer occurs, the problem of body contact becomes a more complicated one. Further examination will focus on the analysis of friction surface of the needle bearing elements in real conditions, in the lubricated environment and under kinematic load.

4. The analysis of friction surfaces of the needle bearing elements

This part gives an analysis of the change in the contact area of elements cooperating under static load in dry environment when two rollers with parallel axes – the needle rollers and the shaft neck – work against each other [16, 17]. Since the nominal contact areas are relatively small – especially when compared with the overall dimensions of the aforementioned elements – pressure in contact areas are significant and accompanied by concentration of stresses. Hertz's solution made it possible to establish the nature of mutual interaction between the bodies, namely the width of contact area and maximum pressure. Next, the area of contact surface was measured, taking into consideration various diameters of the needle and the shaft, various values of Young's modulus and Poisson's ratio and diversified load. The above mentioned will be borne in mind while examining the wear of bearing elements in the energetic perspective. Deformed areas of friction surfaces are characterised by specific features which directly influence the durability of the elements of a friction pair. This, in turn, means that it is essential to choose appropriate friction materials and to analyse their geometry.

4.1. Description of the stand

The examination stand (Fig. 12) consists of an apparatus responsible for loading F the bearing, the electric motor (1) with a fluent regulation of rotation speed 0 – 1500 rpm, the torque meter

(2) responsible for constant measurement of the friction moment, the rotation sensor: the shaft (7), the needle roller around the axis of the shaft (6) and its axis of symmetry (5), the thermo-visional camera (8), the converter processing the above-mentioned parameters (9) and the computer for recording data (10). The needle bearing with a fixed, immobile outer ring is placed on the shaft neck and loaded with radius force *F*. The motor drives the shaft neck up to a certain speed and then the rotation sensors, the torque meter and the camera record the rotation speed of the shaft and the needle, the moment of friction and the heat emitted.

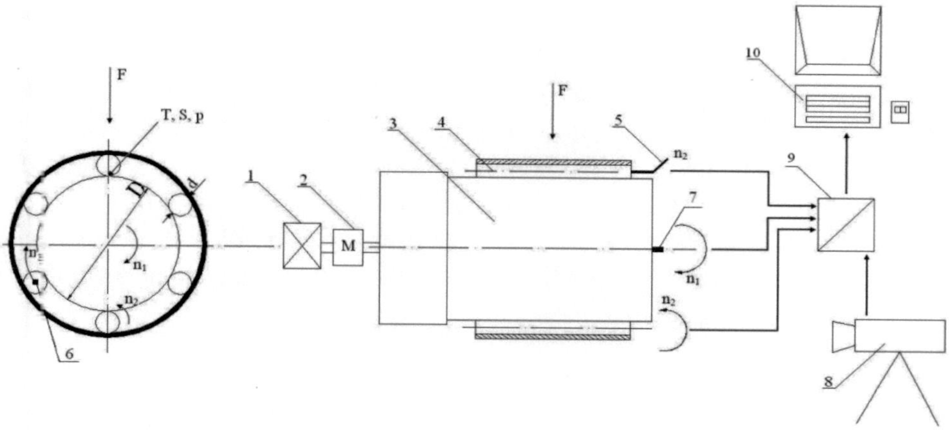

Figure 12. The examination stand: 1 – the source of energy, 2 – the torque meter, 3 – the shaft neck of diameter *D*, 4 – the needle roller of diameter d, 5 – the rotation sensor n2 of the needle against the symmetry axis, 6 – the rotation sensor n3 of the needle around the axis of the roller, 7 – the rotation sensor n1 of the roller, 8 the thermo-visional camera, 9 – the converter, 10 – the computer for processing data

4.2. The analysis of contact areas of friction elements

In the place of contact of two elastic bodies pressed against each other with some force, some contact stresses within a certain field of mutual contact occur. They reach significant values even in the situation when the pressing force is relatively small, which, as a consequence, may lead to exceeding the acceptable limit of the material effort. This is of paramount importance during the work of needle bearings which are under considerable load. Needle bearing requires an adequate positioning of the needles in relation to the shaft neck; in the right position – i.e. when the axes of the elements are parallel – the contact area of these bodies equals the area of the ellipse of the length which is the same as the length of the needle and the width 2*b* calculated by means of Hertz's solution:

$$b = \sqrt{P \frac{D_1 \cdot D_2}{D_1 + D_2} \left(\frac{1 - v_1^2}{E_1} + \frac{1 - v_2^2}{E_2} \right)},$$

where *P* is the strength per each unit of the rollers' length, D_1, D_2 are the diameters of the shaft neck and the needle respectively, v_1,v_2 are Poisson's E_1,E_2 are Young's modulus. The indexes $_{1,2}$ refer to the roller 1 (the shaft neck) and 2 (the needle roller) respectively.

The area of the friction surface of the elements of bearing is influenced not only by the adequate mutual positioning of the cooperating elements, but also by the changeable relation of the diameters of the needle and the roller, changeable values of Young's modulus, Poisson's ratio of the materials used, and the change of load.

The basic theoretical perspective assumed while calculating contact stresses and the width of the contact area between co-working bodies was based on Hertz's theory, drawing on the following premises: the contacting elements are made from a homogenous, isotopic material; they are limited by the smooth surfaces with a regular curvature; and within the contact area some deformations occur (Dietrich, 2003). The analysis of the change in the contact area of the elements of needle bearing presented below was prepared on the basis of all the premises of Hertz's theory concerning the work of two rollers (the shaft neck, the needle roller) with parallel axes, working under static load in dry environment. In the examination of the factors which have an impact on the measure of contact area of the elements cooperating in the form of "the shaft neck – the needle roller" interaction is particularly important as the contact area influences the change in the moment of movement resistance in the process of bearing.

In the first test eight different needles of 1,4 [mm] up to 2,5 [mm] in diameter were juxtaposed with four rollers of 10, 20, 30, 40 [mm] in diameter. The contact area of the cooperating elements was calculated taking as an assumption the constant load of 200 [N/mm]. Different combinations of elements (needles and rollers) were tested. The bigger the diameter of the needle, the bigger the contact area becomes – this tendency can be measured by means of power equations (Fig. 9).

In the case demonstrated above, the biggest changes in the contact area can be observed while putting together the needles of bigger diameters with the given set of rollers, i.e. for the needle of 1,4 [mm] in diameter cooperating with the rollers of 10 and 40 [mm] in diameter the contact area increased by approximately 5 [%], while for the needle of 2,5 [mm] in diameter cooperating with the same rollers the contact area increased by approximately 9 [%].

Fig. 13. shows the influence of Young's modulus E, Poisson's ratio υ and the roller diameter D on the area S under a constant load. The range of Young's modulus assumed in the test was identical with the one meant for the materials used in the construction of shafts and bearings. The increase in the contact area of cooperating elements when E and υ are changed is not significant; in the situation illustrated in Fig. 13, the increase approximates to mere 0.8 [%] when E is doubled. This is connected with the assumed load which is transmitted onto the whole length of the needle (the axes of the shaft and the needle roller are parallel). In case of materials that are characterized by Young's modulus of 200 [GPa] and more, the assumed load of 200 [N/mm] does not enforce any deformation in the cooperating elements, i.e. the needle of 2 [mm] in diameter and a set of rollers of 20, 30, 40 [mm] in diameter.

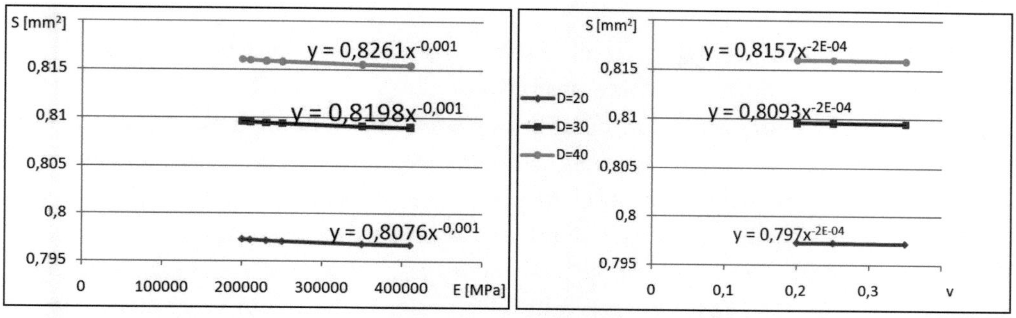

Figure 13. The influence of Young's modulus – E of the material, Poisson's ratio – v, the diameter of the shaft neck (D) on the contact area of the elements of bearing (S)

Assuming that the task of loading is an indispensable factor in creating stresses and deformations in the needle roller and the shaft neck, the influence of radius strength P on the contact area becomes obvious.

From the analysis of the diagram in Fig. 13 it is clear that the increase in the area S is directly proportional to the growing strength P. In the test a needle roller of 2 [mm] in diameter (E = 210 [GPa], v = 0.3) was used as well as a set of four rollers of 10, 20, 30, 40 [mm] (E = 200 [GPa], v = 0.25) in diameter. The diagram of the increase in the contact area with the load of bearing by radius force ranging from 200 to 450 [N/mm] is reflected by power function (Fig. 14) which is connected z hardening of the contact area under the influence of growing force.

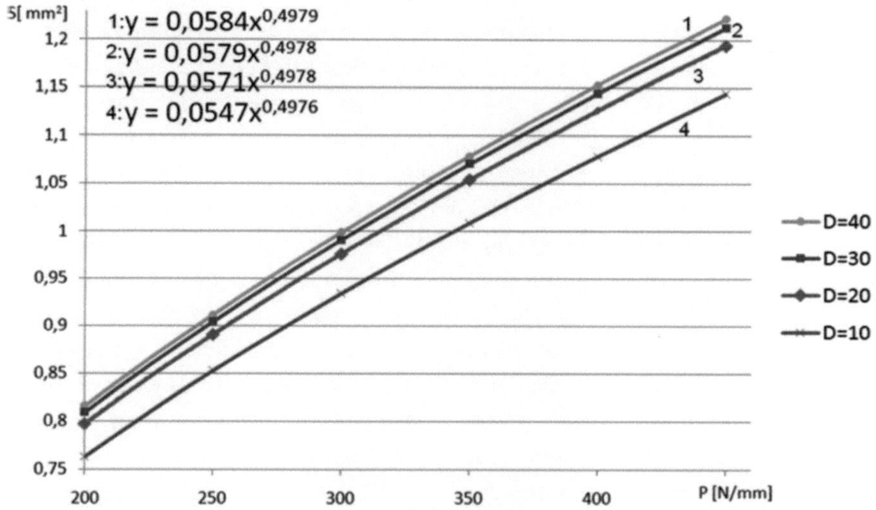

Figure 14. The influence of load (P) and the diameter of the shaft neck (D) on the contact area of the elements of bearing (S)

To model the process of friction, to calculate the rotational speed of the shaft and the needle roller, and to estimate the importance of slide resistance in the overall process of friction, the following examination stand was used [18,19].

5. Summary

The work refers to different aspects of research methodology and the difficulties in the interpretation of the durability and reliability of the bearings. The proposition of own methods of research and empirical descriptions was done. The material presented in the work was divided into four separate subsections: on research methodology, work, geometry, and the impact of these elements on the friction and wear. All tests were performed for three replicates at each point. The investigations were conducted with using different tribological testers and tribological stands. Should add that all of the test, later in this chapter bearings are factory items and their designations are in accordance with PN, PN-EN and ISO standards. All tests were performed for three replicates at each point.

On the basis of the test results the following conclusions have been made:

The results of tribological tests indicate different friction characteristics of the evaluated elements. Bigger resistance to motion is noticed at the beginning of the test. Afterwards it becomes stable – at minimal level, at the end of the process. The data included in this section indicate entirely different test results obtained with two different methods. This may lead to totally different conclusions concerning the tribological properties of tested elements. The data presented in fig. 2, 3 and 4 clearly indicate the lack of correlation between MB-1 and MB-2 methods applied in the tests. This means that conducting a scientifically reliable evaluation of tribological properties of such elements requires applying one method. However, the results obtained with this method may not reflect the actual friction nodes when one is unable to faithfully reproduce their actual working conditions. The conducted examination and simulations have demonstrated the changes in the contact area of friction pair elements when two rollers with parallel axes (i.e. the needle roller and the shaft neck) press against each other. Examination of the factors that can have an influence on the contact area of cooperating elements, i.e.: changes in the needle roller diameter and the change in the shaft diameter, are of particular importance because the contact area alters the resistance moment of motion in the bearing. The suggested mathematical model and the examination on a stand help to estimate the extent to which the bearing friction and the sliding friction matter in the overall balance of motion resistance. While examining the shaft neck wear in the function of initial slackness, the conditions of movement from the normal wear to the pathological one were established by means of setting up the critical parameters: the slackness, the intensity of wear in the function of load. It has been observed that the movement from the normal wear to adhesion is characterized by a gradual increase of the intensity of the elements wear. The conducted examination and simulations have demonstrated the changes in the contact area of friction pair elements when two rollers with parallel axes (i.e. the needle roller and the shaft neck) press against each other. Examination of the factors that can have an influence on the contact area of cooperating elements, i.e.: changes in the needle roller diameter and the change in the shaft diameter, changes in the value of Young's modulus, Poisson's ratio of the materials used and changes of load are of particular importance because the contact area alters the resistance moment of movement in bearing. The analysis of the above-mentioned relations in the energetic perspective made it possible to conclude that deformed areas of friction surfaces are characterized by specific properties which influence the durability of the elements of the friction pair. This means that the materials meant for friction pairs should be properly selected and their geometry carefully analyzed. While examining the shaft neck wear in the function of initial clearance, the conditions of movement from the normal wear to the pathological one were established by means of setting up the critical parameters: the clearance, the intensity of wear in the function of load. It has been observed that the movement from the normal wear to adhesion is characterized by a gradual increase of the intensity of the elements wear. The examination was carried out for the nominal clearance of 0.065 mm and other clearances achieved as a sum of the nominal one and the value of reduction of the shaft neck diameter. The associations received in this way made it possible to examine the process of friction with clearances of 0,1, 0,2, 0,3, 0,4, 0,5 [mm] respectively. A hundred-hour examination was carried out for each value of clearance in bearing; next, the micro-structure of the friction surface was examined under an optic microscope with magnification ×300, which lead to the distinction of three different surfaces of wear characteristic for normal, mechano-chemical and pathological wearing (Fig. 10 b). On the basis of the results received is can established that the enlargement of clearance between the elements in the friction pair leads to a considerable enlargement of the intensity in the shaft neck wear, which, in turn, results in the worse durability of the connection. Fig. 10 a shows the wear of the shaft neck understood as function of the initial clearance. Point A as a critical clearance with which the friction pair works in the conditions of normal wear was achieved as a result of crossing the tangents led to the curve of wear in areas I and III (Nachimowicz et al., 2007). In real conditions, when the surfaces of the bodies in direct contact are uneven, and anisotropy of top layer occurs, the problem of body contact becomes a more complicated one. Further examination will focus on the analysis of friction surface of the needle bearing elements in real conditions, in the lubricated environment and under kinematic load.

Author details

Jerzy Nachimowicz
Department of Building and Exploitation of Machines,
Mechanical Faculty, Bialystok University of Technology, Bialystok, Poland

Marek Jałbrzykowski*
Department of Materials and Biomedical Engineering,
Mechanical Faculty, Bialystok University of Technology, Bialystok, Poland

6. References

[1] Szczerek M.: The methodological problems of systematization of experimental tribological research, Publisher ITE Radom, 1997.
[2] Budzoń P., Lenkiewicz W., Olesiak Z.: Standardization of documentation of tribological research on the example of rolling friction, The Problems of machine Exploitation 1, 1986, 233-240.

---

* Corresponding Author

[3] Sosnovskiy L.A., Koreshkov V.N., Yelovoy O.M.: Methods and machines for wear-fatigue tests of materials and their standardization, Proc. I[st] World Tribology Congress Mechanical Engineering Publication Ltd. London, 1997, 723.

[4] Styp-Rekowski M.: The importance of design features for durability pitched ball, Publisher Institutional University of Technology and Agriculture in Bydgoszcz, 2001.

[5] Погодаев Л.И., Кузьмин В.Н., Дудко П.П.: Повышение надежности трибосопряжений. Санкт-Петербург: Академия транспорта Российской Федерации.- 2001, с. 316.

[6] Нахимович Е.: Применение наплавок для восстановления изношенных деталей. /Труды первого международного симпозиума по транспортной триботехнике «Транстрибо-2001». – СПб.: Изд-во СПбГТУ, 2001,с.140-143.

[7] Nahimovich E., Kaczynski R.,(2003), On the criteria of the influence of the lubrication medium on the durability of bal bearings, *Mechanical Engineering,* Vol. 12, p. 49-52.

[8] Nahimovich E., Chulkin S., (2002), The comprehensive approach to problem solving on a heigtening of longevity both endurance of materials and elements of machines, *Mechanical Engineering,* Vol. 4, p. 42-44.

[9] Nachimowicz J,: Tribological characteristics of bearing system of road wheels, International Conference "Friction 2004". Modelling and simulation of the friction phenomena in the physical and technical systems. Warsaw University of Technology 2004, p. 59-63.

[10] Нахимович Е., Погодаев.Л.И. : Моделирование процесса изнашивания и прогнозирование долговечности опор качения, Изд. СПбГТУ, Санкт-Петербург, 2002, с.129.

[11] Dietrich M. (red.): Fundamentals of Machine Design, Publishing and Scientific - Technical Ed. 3. T.2, Warsaw, 2003, p. 391 – 400.

[12] Hebda M ., Wachal A.: Tribology, Publishing and Scientific - Technical Ed., Warsaw, 1980, p. 145-147, 443-446.

[13] Nachimowicz J., Baranowski T., Jabłoński D.: The tribological aspects of the work needle bearing, XXIII Symposium of Machine Construction, Rzeszow – Przemysl 2007, p. 109 – 115.

[14] Nachimowicz J., Jałbrzykowski M.: Methodological Aspects of Evaluating Tribological Properties of Ball Bearings, Solid State Phenomena Volumes, Mechatronic System and Materials III, Trans Tech Publications, Switzerland, p. 147-149.

[15] Totten G., Liang H.: Mechanical Tribology, Marcel Dekker, New York 2004, p.64-69, 199-205.

[16] Nachimowicz, J., Baranowski, T., Jabłoński, D.: Tribological aspects of the work bearing needle, XXIII Symposium of Machine Construction, Rzeszow-Przemysl, 2007, p. 109–115.

[17] Nachimowicz, J., Jałbrzykowski, M.: Methodological Aspects of Evaluating Tribological Properties of Ball Bearings, Solid State Phenomena Volumes, Mechatronic System and Materials III, Trans Tech Publications, Switzerland, 2009, p. 147–149.

[18] Dietrich, M.: Fundamentals of Machine Design, Wyd. 3, T. 2, Warsaw, 2003. p. 391–400.

[19] Myshkin, N.K., Petrokovets, M.I.: Трение, Смазка, Износ, Moscow, 2007. с. 205–208.

Chapter 3

Radial Ball Bearings with Angular Contact in Machine Tools

Ľubomír Šooš

Additional information is available at the end of the chapter

http://dx.doi.org/10.5772/51004

1. Introduction

The decisive criteria of the quality of machining tools are their productivity and working accuracy.

One innovated method for improving the technological parameters of manufacturing machines (machine tools) is to optimise the structure of their nodal points and machine components.

Because of the demands on machine tool productivity and accuracy, the spindle-housing system is the heart of the machine tool, Figure 1, [1]. Radial ball bearings with angular contact are employed in ever increasing arrays. The number of headstocks supported on ball bearings with angular contact is increasing proportionally with the increasing demands on the quality of the machine tool [2]. This is because these bearings can be arranged in various combinations to create bearing arrangements which can enable the reduction of both radial and axial loads. The possibility of varying the number of bearings, their preload value, dimensions and the contact angle of bearings used in the bearing nodes, creates a broad spectrum of combinations which enable us to achieve the adequate stiffness and high speed capabilities of the **Spindle-Bearings System** (SBS) [2], [3]. Adequate stiffness and revolving speed of the headstock are necessary conditions for meeting the manufacturing precision quality and machine tool productivity required by industry.

When designing a machine tool headstock, the starting point is the design of the spindle support, as this limits the stability, accuracy and production capacity of the machine by its stiffness and revolving speed. However, the parameters influencing the stiffness and frequency can *act in opposition to each other*. The selection of the type of bearing has to take into consideration the optimization of its stiffness and revolving speed characteristics. The maximum turning speed of the bearings is a function of the maximum revolving speed of the individual bearings, their number, pre-load magnitude, manufacturing precision, and the types of lubrication used.

The stiffness of the SBS depends on the stiffness of the bearings and the spindle itself. There are several methods that can be employed for determining the static stiffness of the spindle system, e.g. [1] and [2].

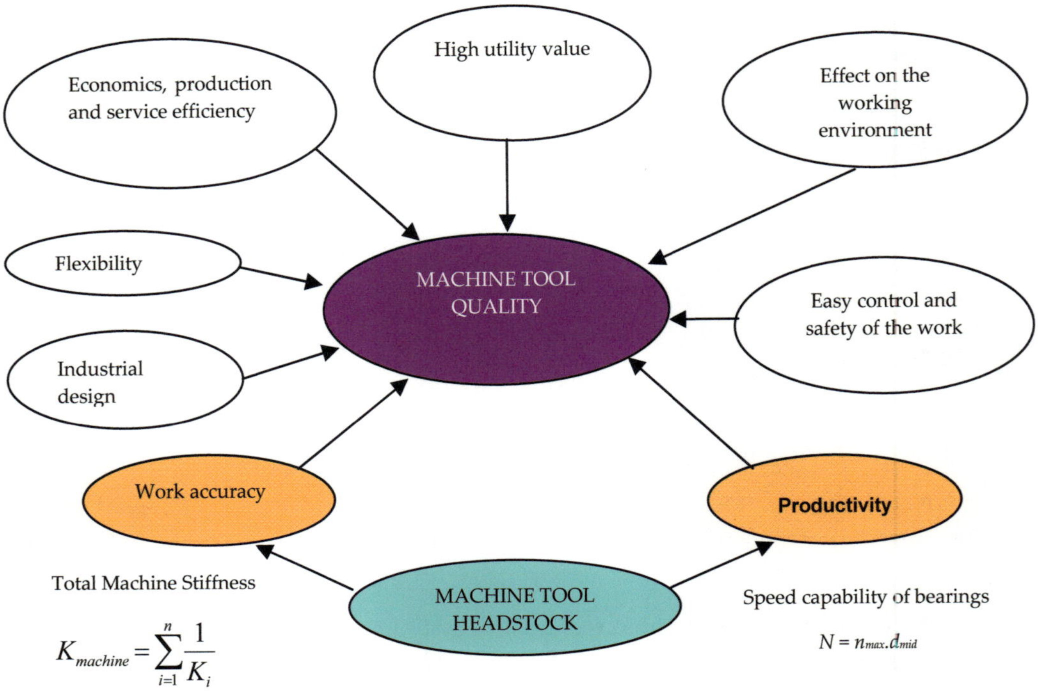

Figure 1. Factors influencing the Quality of Machine Tools, [4].

However, one problem which has not yet been solved is the calculation of the stiffness of the bearings, (or nodes of bearings) in the individual housing. Accurate calculation of the stiffness of the bearing nodes requires the determination of the static parameters of each bearing. From a mathematical point of view, this can be solved by using a system of non-linear differential equations, which requires the use of computers. To simplify the design, we need a static analysis which provides the basis for the dynamic characteristics of the mounting, and of the machine itself. Designers often prefer the conventional and proven methods of mounting, without taking into account the technical and technological parameters of the machine.

For the design engineer, it is important to be able to undertake a quick evaluation of various SBS variants at the preliminary design stage. The success of the design will depend on the correct choice of suitable criteria for the SBS, and if the design engineer has adequate experience in this field.

1.1. Headstock – The heart of the machine tool

The headstock, whether tool or workpiece carrier, has a direct influence on the static and dynamic properties of the cutting process, Figure 2. The spindle-bearing system (SBS) stiffness affects the surface quality, profile, and dimensional accuracy of the parts produced. It also has a direct influence on machine tool productivity because the width of cut influences the initiation of self-induced vibration; it is directly proportional to machine tool stiffness and damping.

Complex analysis of the SBS is very difficult and complicated, [5]. The analysis requires an advanced understanding of mathematics, mechanics, machine parts, elasto-hydrodynamic theory, rolling housing techniques, and also programming skills. The results of our research into SBS have been divided into three parts:

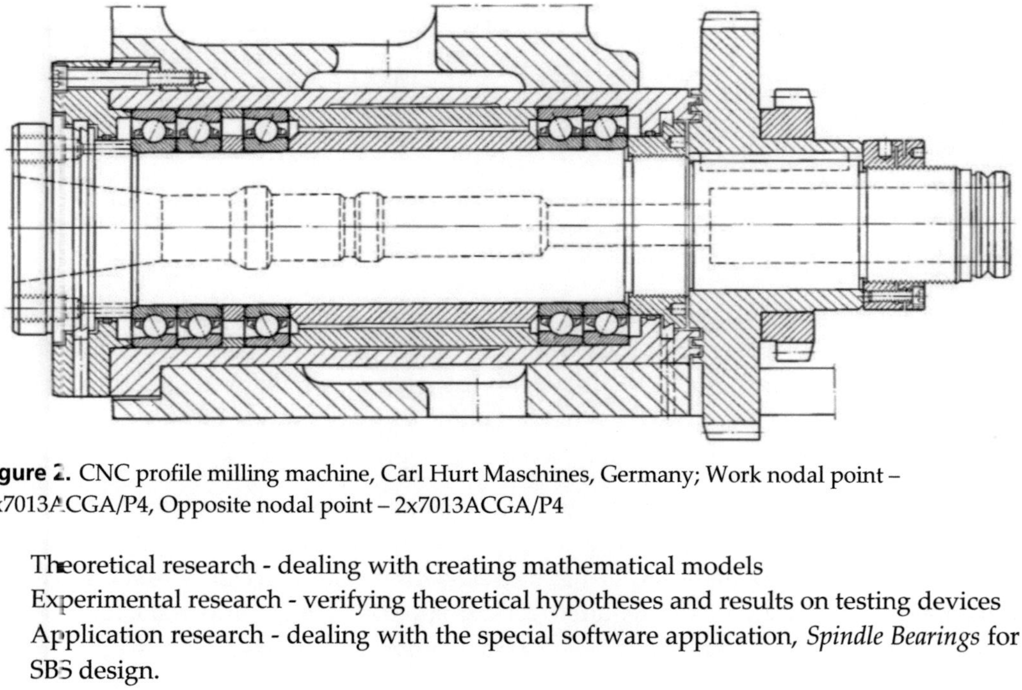

Figure 2. CNC profile milling machine, Carl Hurt Maschines, Germany; Work nodal point – 3x7013ACGA/P4, Opposite nodal point – 2x7013ACGA/P4

- Theoretical research - dealing with creating mathematical models
- Experimental research - verifying theoretical hypotheses and results on testing devices
- Application research - dealing with the special software application, *Spindle Bearings* for SBS design.

2. Theoretical research

A modular structure of the theoretical research is shown in Figure 3.

2.1. Primary static analysis

2.1.1. Speed

The productivity of a machine tool, (Figure 1) can be increased in at least two different ways:

1. Externally - by shortening working time - within a working cycle
2. Internally - by reducing machining times (increasing the cutting width) - technological issues.

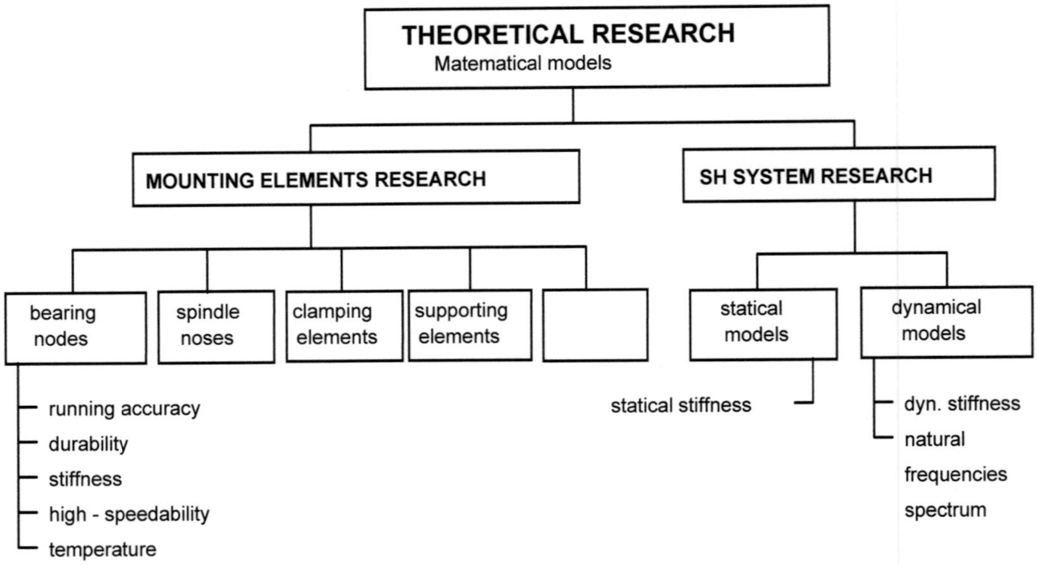

Figure 3. Modular structure of theoretical research, [5]

The philosophy of intelligent manufacturing systems applied to production processes minimise lost time. Further reducing lost time is expensive and has limited effectiveness at current levels of technological development. It has been shown that increased productivity can be achieved for example by changing the cutting speed. However this has a direct effect on tool life and on the dynamic stability of the cutting process.

The cutting speeds in machining processes depend on the technology applied, the cutting tool, and the workpiece material. The cutting speed also relates directly to the high-speed capability, and average diameter, of the bearings, the so-called factor $N = n_{max}.d_{mid}$.. Thus, from the point of view of the required cutting speed, the most important factor is the revolving frequency capacity of a spindle which is supported on a bearing system.

The calculation of the headstock's maximum revolving speed is relatively simple. The highest revolving speed of a bearing node is calculated on the basis of the highest revolving speed of one bearing, multiplied by various coefficients reflecting the influence on the bearings, the bearing arrangement, bearing precision, their preloaded value, and lubrication and cooling conditions.

2.1.2. Stiffness

The total static stiffness of machine tools is, in almost all cases, limited by the stiffness of the weakest parts. Amongst all the elements, the Spindle-Bearings System of the machine tool plays the most important role.

From results of structural analyses, the headstock can be considered as the heart of the whole machine tool. The design and quality of the machine tool must respect the quality of the drives and their features.

The headstock (as tool, or workpiece carrier), has a direct influence on the static and dynamic properties of the cutting process. The Spindle-Bearing System's stiffness also influences the final surface quality, profile, and dimensional accuracy of the workpiece.

The problem here is how quickly the headstock stiffness can be calculated with sufficient precision. The headstock stiffness must be calculated according to the deflection at the front end of the spindle, because the deflection at this point directly affects the precision of the finished product. The deflection at the spindle front end is the accumulation of various other, more or less important, partial distortions. The radial headstock stiffness can be calculated as follows:

$$K_{rc} = \frac{F_r}{y_{rc}} \tag{1}$$

The individual headstock parts, (spindle and bearing arrangement), create a serial spring arrangement and it is evident that the resulting stiffness K_{rc} is limited by the stiffness of the weakest part. An expert can see which part should be improved, and which partial distortions need to be minimized.

2.2. Simplified method of calculation

The calculation of spindle front end deflection, which takes into consideration all the important parameters, can only be achieved by using powerful computers. The analysis can be carried out by standard or custom software programs.

Calculating the many combinations of SBS arrangements is very demanding on time and money. Undertaking stiffness analysis using standard programs depends on the engineers' experience. The results can be open to questionable even when a suitable mathematical method is used (finite element method, boundary element method, Castilian's theorem, graphic Mohr's method, etc). This is because the headstock box, bearings or bearings nodes are statically indefinite systems which produce a nonlinear deformation of the node when under load.

Special software programmes are very expensive. They are developed using the most up-to-date theoretical and practical knowledge. These programs have been developed by research institutions and bearing producers and the possibility of using such programs significantly influences their position on the SBS market. Taking the above into account, engineers would benefit from the existence of a simplified method of static analysis. Such a methodology would enable the engineer at the preliminary design stage to limit the number of possible spindle-bearing variants and determine the direction which would lead to the optimal SBS design, [6].

The main methodological advantage of computer analysis is the possibility of repeating single calculating algorithms in a matrix shape. To this end a special software package, *"Spindle Headstock"* was developed at the Department of Production Engineering in the Faculty of Mechanical Engineering at STU, Bratislava, [7].

The resulting radial deformation, y_{rc}, of the front spindle end is shown in Figure 4.

Resulting static distortion of the front-end spindle equals

$$y_{rc} = y_0 + y_1 + y_t + y_a + y_v + y_{sb} + y_h \tag{2}$$

Our experience has shown that whatever mathematical method and software is used, the spindle distortion caused by *bending moments* y_0 and by *bearing compliances* y_l have the greatest influence on the resulting front end spindle distortion, [6].

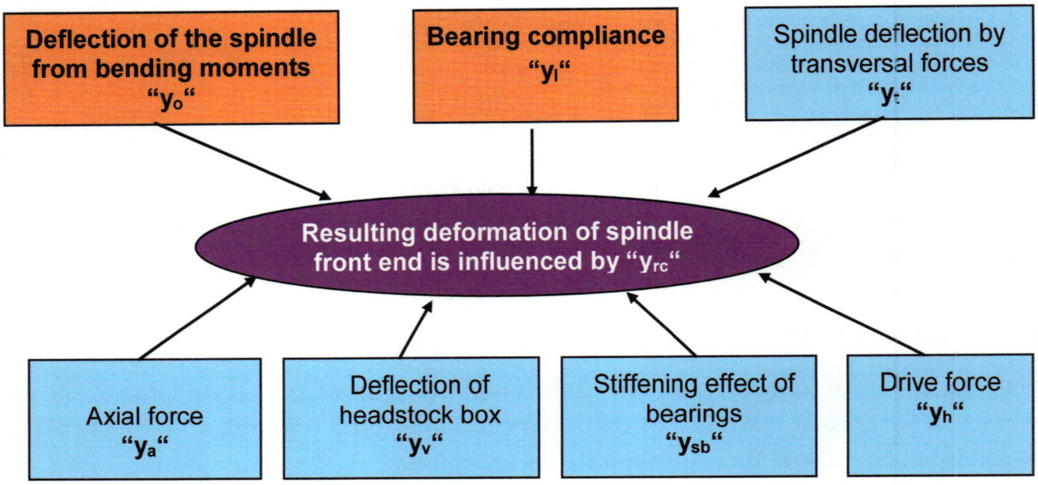

Figure 4. Factors influencing the resulting deflection

Then

$$y_{rc} = y_0 + y_1 \tag{3}$$

where the distortion caused by bending moments is as follows:

$$y_o = \frac{F_r a^2}{3E}\left[\frac{a}{J_a} + \frac{L}{J_L}\right] \tag{4}$$

and the deflection caused by bearing compliance is as follows:

$$y_l = \frac{F_r}{L^2}\left[\frac{a^2}{K_B} + \frac{(L+a)^2}{K_A}\right] \tag{5}$$

Increasing moments of inertia "J_a", "J_L" were calculated as follows:

$$J_a = \frac{\pi}{64}\left[D_a^4 - d_a^4\right] \text{ and } J_L = \frac{\pi}{64}\left[D_L^4 - d_L^4\right] \quad (6)$$

The definition of the quantities is shown in Figure 5. The individual headstock parts (spindle, bearing arrangement,) create a serial spring arrangement, and it is evident that the resulting stiffness K_{rc} is limited by the stiffness of the weakest part, [1] [2] and [9].

At the same time, parameters "F_r", "a", "L" influence the value of both deflections. The spindle deflection caused by bending moments can be decreased by the following methods:

- increasing the material modulus of elasticity "E",
- increasing moments of inertia "J_a", "J_L" by a change of spindle diameters "D_a", "D_L", "d_a", "d_L".

The resulting static distortion of the spindle front-end can be explicitly described by a multi-parametrical equation in the form of:

$$y_F = f\,[E, F_r, a, L, J_a,, (D_a, d_a)\ J_L(D_L, d_L), K_A, K_{B,}\rho] \quad (7)$$

- spindle material and dimensions (E, D_a, d_a, D_L, d_L)
- loading forces position, orientation and magnitude (F_r, N, r_F, b)
- bearing arrangement configuration and stiffness (K_A, $K_{B,}$)
- spindle and bearing arrangement space configuration (L, a)
- spindle box construction (k_ξ, ρ)

There remains one significant problem with the calculation of the bearing nodes, and that is that they are statically indeterminate systems.

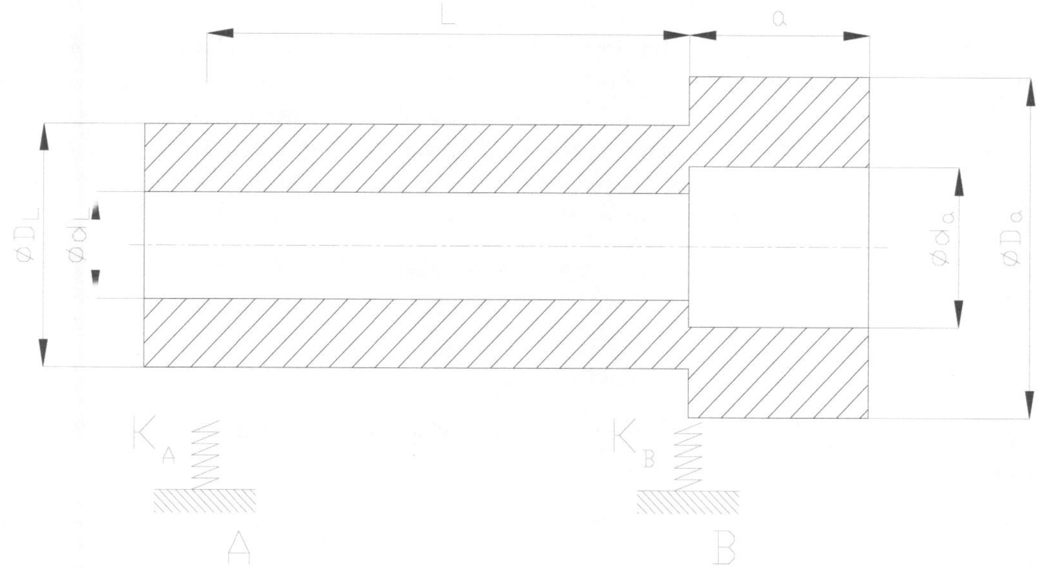

Figure 5 Cross section scheme of the spindle-bearing system

2.3. Dynamic analysis

While static analysis of *SBS* describes spindle behaviour in a static mode, dynamic analysis describes SBS behaviour under real conditions, in real running time, and so the real operational state is better represented. It is very important to know the dynamic characteristics, especially in high-speed headstocks. It is important to ensure that the operational revolving frequencies do not fall within the resonant zone. When this happens, the vibration amplitude of the spindle is considerably increased, and the spindle's total stiffness falls to unacceptable levels.

The most common determining dynamic characteristics of SBS are:

- the spectrum of natural (resonant) frequencies (usually the first three frequencies),
- the amplitudes of vibrations along the spindle independent of the revolving frequencies of the spindle,
- the resonant amplitudes of vibrations,
- the dynamic stiffness of the spindle (at the given speed of the spindle).

The SBS dynamic properties (dynamic deflection of spindle front-end, natural frequencies spectrum) [5], are affected by factors shown in Figure 6.

Mathematical models for determining the dynamic properties of a spindle

Currently, the only reliable method for determining dynamic properties is to use experimental measurements. Therefore it is very useful to create reliable mathematical models for determining these dynamic properties.

In line with spindle mass reduction, mathematical models are divided into:

1/ discrete with 1º, 2º and Nº degrees of freedom,
2/ continuous.

The discrete mathematical model developed for measuring the revolving vibration of spindles with N° degrees of freedom is worked out in [1], [5]. This mathematical model for calculating the dynamic properties of the spindle enables us to include in our calculation the effects of the materials, the dimensions of the rotating parts, the bearing node stiffness, and the radial forces generated by the cutting process and drive. The results calculated reflect a spectrum of natural frequencies and the dynamic deflection of the spindle under discrete masses.

The deflection of spindle y_i loaded with concentrated forces at the ith point can be expressed in the form:

$$yi = a_{i1}F_{1o} + a_{i2}F_{2o} + \ldots + a_{ik}F_{ko} + \ldots + a_{in}F_{no} \ (m) \tag{8}$$

where a_{ik} (m/N) is Maxwell's affecting factor. Every mass point on the spindle produces centrifugal force

$$F_{io} = m_i y_i \omega^2 \quad (N) \tag{9}$$

where m_i (kg) is mass ith discrete segment.

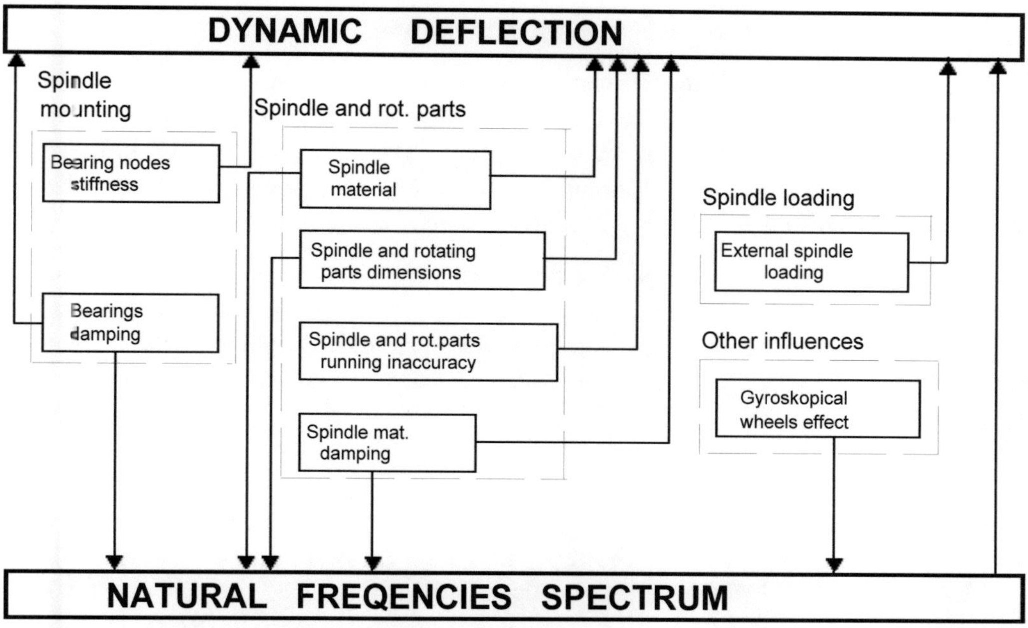

Figure 6. Factors affected by SBS dynamic properties

The application of the aforementioned equations and their modification for masses of "n" value, will create a system of homogenised algebraic equations where the results of determinant D are angular natural frequencies of the transverse vibrations of the spindle ω_i (rad.s^{-1}).

The procedure for determining the dynamic deflections y_i, when the dimensions of the spindle, rotating parts, stiffness of bearing arrangement, and the external radial forces are taken into consideration, is very similar to the previous one. These procedures are described in [6].

$$\Delta = \begin{vmatrix} 1-a_{11}\, m_1\, \omega^2 & -a_{12}\, m_2\, \omega^2 & \dots & -a_{1n}\, m_n\, \omega^2 \\ -a_{21}\, m_1\, \omega^2 & 1-a_{22}\, m_2\, \omega^2 & \dots & -a_{2n}\, m_n\, \omega^2 \\ \dots & \dots & \dots & \dots \\ -a_{n1}\, m_1\, \omega^2 & -a_{n2}\, m_2\, \omega^2 & \dots & 1-a_{nn}\, m_n\, \omega^2 \end{vmatrix} = 0 \qquad (10)$$

It is relatively easy to transform this mathematical model into a computer readable format and the calculation of the dynamic characteristics can be quickly achieved.

2.4. Theoretical research on bearing nodes

2.4.1. Arrangements of nodal points

Usually, radial ball bearings with angular contact arrangements in their nodal points contain 2, 3 or more bearings, see Figure 7.

2.4.2. Criteria for selecting the arrangement of bearings

The number of spindle bearing systems supported on ball bearings with angular contact increases proportionally with increasing demand on the machine tool. By varying the bearings and their arrangement in the bearing nodes (DB, DF, DT, TBT,TTF, QBC, ..), the value of the contact angle, magnitude of preload, and type of flanges can be optimized to suit the required, resulting stiffness and speed-capability of the spindle-bearing system.

In order to assess the maximum permissible speed of different types of spindle rotations, a parameter for the so-called high-speed characteristics has been introduced: $N = n_{max}.d_{mid}$, where "n_{max}" denotes the maximum spindle revolutions and "d_{mid}" the medial diameter of the bearings. Following this parameter, roller bearings of machine tool spindles can be divided into 3 basic groups, [10]:

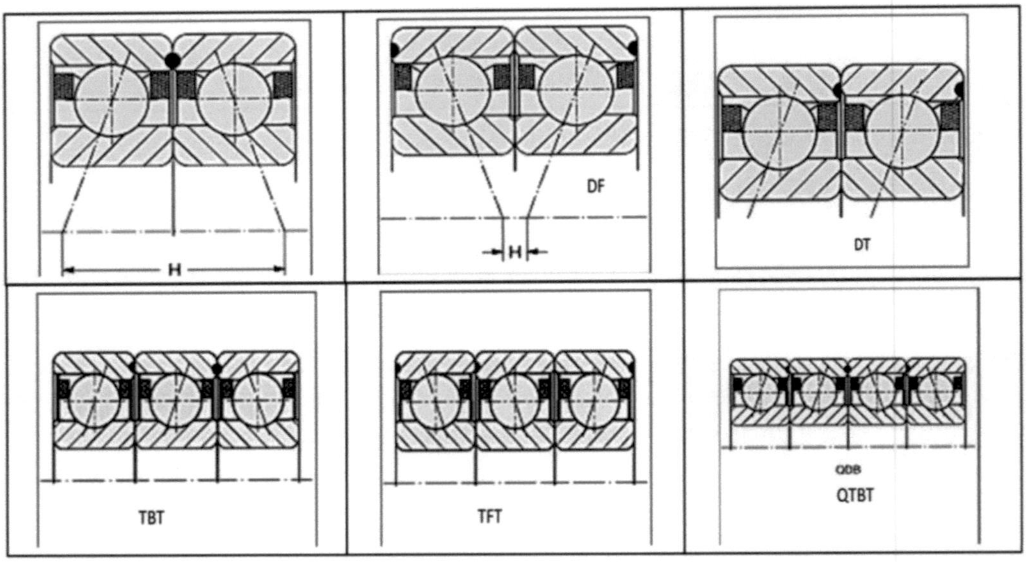

Figure 7. Arrangements in nodal points

GROUP 1: N = (0,1 – 0,5) .10^6: Headstocks of heavy duty machines for turning, milling and drilling operations. In these machines the spindles are predominantly mounted using double-row roller bearings in combination with axial ball bearings, or tapered bearings. We can assume that the linearization of the deformation curve in roller bearings is sufficiently accurate, which simplifies the calculation of the radial stiffness of the nodal point, [2]. These *mountings* offer high stiffness and load-bearing capacity and quiet operation.

GROUP 2: N = (0.4 – 1) .10^6 is characterized by bearings of medium size and are found in smaller NC and CNC turning, milling, drilling and grinding machine tools. The maximum possible speed in bearings with linear joints is limited by the heat produced in the head, and therefore they are used only for the mounting of spindles with the lowest values of coefficient N. Developments in the field of increased speed capability is focused on bearings with point contacts, as these have better friction characteristics.

GROUP 3: Spindles mounted in bevelled radial bearings with optimized structure (design) and using new composite materials enabling high-speed operation, N = (0.8 – 2.5) x 10^{6}, which is typical for high-speed machining.

Spindle mountings using only radial bevelled bearings, (table 1), [11] can be divided into 2 basic types:

- spindles mounted on bearing nodes with "directionally" arranged bearings, with equal orientation of contact angles in each nodal point 1, 2, 3, and 7, table 1.
- spindles mounted on nodal points with bearings arranged according to shape. Bearings are arranged in "O" or (X) shape, in combination with "T".

A typical feature of the nodes of spindle bearings is the application of pre-stressing, which provides the stiffness of the nodal point and reduces any skidding of the rollers at high revolutions.

Pre-stressing can be achieved through three flange design principles:

a. Sprung flange: thermal expansion (dilatation) is eliminated by changing the length of the elastic materials positioned between the flange and the bearings, which ensures minimum change in the pre-stress value.
b. Stiff (Rigid) flange: provided by a fixing nut or casing. This design provides better stiffness characteristics. The pre-stress value is changed due to the influence of thermal dilatation.
c. Controllable flange: axially adjustable (by means of hydraulics), which ensures the required pre-stressing for different operational conditions.

The highest values of the coefficient N can be achieved by using spindles mounted on nodes with a "directional" arrangement of bearings, 1, 2 and 3. When used in conjunction with the controllable flange, the correct types of lubrication and cooling, speeds which are comparable with the maximum revolutions of the bearings themselves can be achieved. Thus they can be applied in high-speed machining [11]. These mounting types, in combination with the sprung support, are mostly used for grinding.

For difficult technological operations requiring considerably higher stiffness in the radial and axial directions, nodal points with bearings arranged according to shape, together with fixed supports are typical.

There is negligible use of hybrids of the basic types of mounting (mounting 5), as shown in table 1. In such cases one nodal point has bearings arranged according to shape, while the other has directionally arranged bearings, (Figure 2). The pre-stressing in the front nodal point is ensured by a stiff flange, and in the rear nodal point by a sprung flange.

Seq. No.	CONFIGURATION Rear bearing node	CONFIGURATION Forward bearing node	$N = n_{max}.d_{mid}.10^6$ $[mm.min^{-1}]$	Characteristic	Use
1.	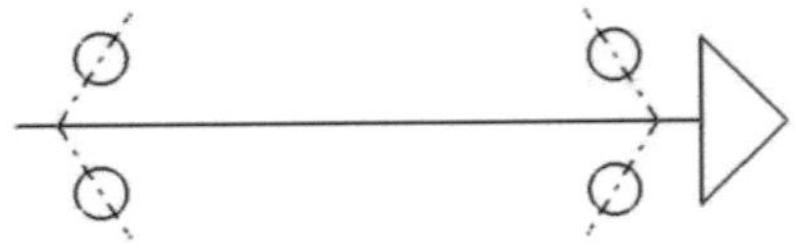 $t_1=1, t_2=0$	$t_1=0, t_2=1$	1,2 -2,5	- single direction of rotation - light axial and radial loads	- grinding internal holes
2.	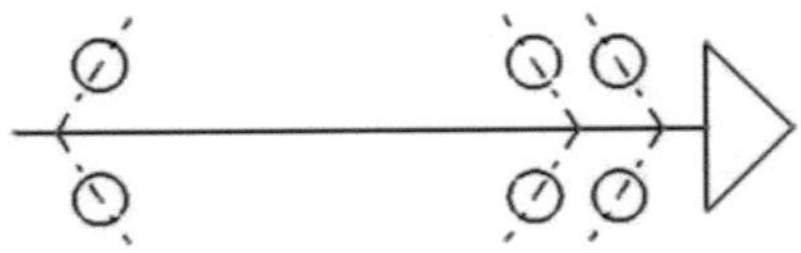 $t_1=1, t_2=0$	$t_1=0, t_2=2$	0,8 – 1,6	- suitable for extremely short spindles - medium axial loads	- finishing machines - drilling of deep holes
3.	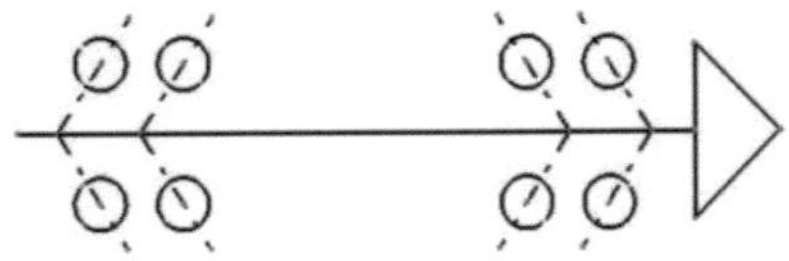 $t_1=2, t_2=0$	$t_1=0, t_2=2$	0,8 -1,4	- medium radial loads - very common method of use	- grinding internal holes - milling - drilling
4.	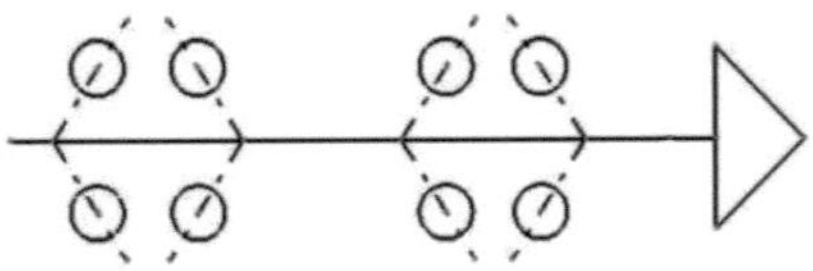 $t_1=1, t_2=1$	$t_1=1, t_2=1$	0,6 – 1	- machining light metals - medium radial loads	- grinding - precision drilling - turning/ lathe
5.	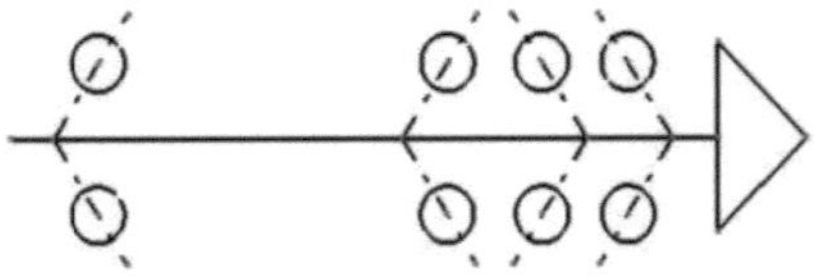 $t_1=1, t_2=0$	$t_1=1, t_2=2$	0,5 – 0,9	- medium axial loads	- drilling of deep holes - milling
6.	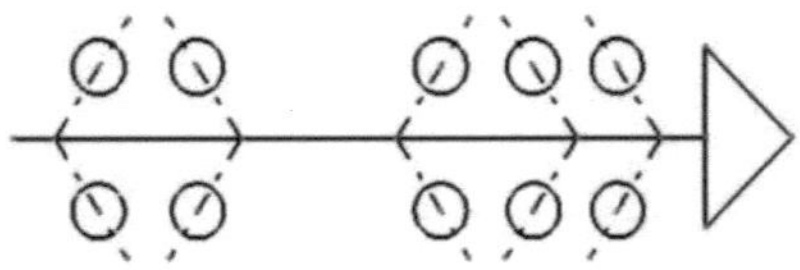 $t_1=1, t_2=1$	$t_1=1, t_2=2$	0,4 – 0,9	- medium axial loads - very common method of use	- turning/ lathe - drilling
7.	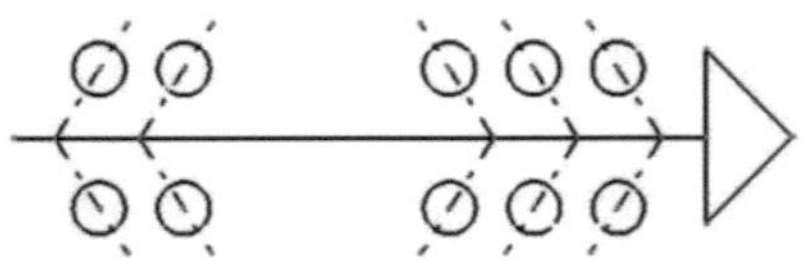 $t_1=2, t_2=0$	$t_1=3, t_2=0$	0,3 – 0,6	- high axial loads medium radial loads	- milling - boring

Table 1. Type of SBS using radial ball bearings with angular contact [11]

2.4.3. *Stiffness*

The stiffness of the bearing arrangement *(K_A, K_B)* is the specific parameter which influences the consequent spindle distortion. We have developed a simplified mathematical model for calculating radial and axial stiffness, [11], [12].

3. The calculation of radial stiffness of nodal points

3.1. Assumptions of solution

According to the Hertz assumptions [13], [14], there is a dependence between load "*P*" and deformation "*δ*" at the contact point of the ball with the plane, given by the relationship

$$P = k_{\delta}.\delta^{3/2} \tag{11}$$

a. the bearings in the nodal points are of the same type and dimensions, with precise geometric dimensions
b. the value of the contact angle is the same for all directionally-arranged bearings in the nodal point, which delivers equal distribution of strain on these bearings
c. radial load is equally distributed onto all the bearings in the nodal point

3.2. Stiffness of nodal points with directionally-arranged bearings

The calculation of the stiffness of a nodal point is based on the stiffness of the bearing itself [15], which is defined as:

$$K_{r1} = \frac{d\ F_{r1}}{d\ \delta_{r0}} \tag{12}$$

As radial displacement δ_{r0} is a function of contact deformation δ_0 of the ball with the highest load [13], the equation for calculating the stiffness of bevelled radial bearings will have the form:

$$K_{r1} = \frac{d\ F_{r1}}{d\ \delta_0}.\frac{d\ \delta_0}{d\ \delta_{r0}} \tag{13}$$

When calculating stiffness, the distribution of load among the rollers must be determined, and the dependence between the load on the top ball and external load must be found. The distribution of load in the bearing can be derived from the balance under static conditions [14],

$$F_{r1} = \frac{F_{r1}}{i} = \sum_{j=0}^{z} P_j.\cos(\alpha_j).\cos(j.\gamma) \tag{14}$$

where $\gamma = \frac{360}{z}$ is the spacing angle of the balls.

The values of contact deformations δ_j and angles α_j differ from each other around the circumference of the bearing and can be expressed as follows, (Figure 8).

$$\delta_j = l_{rj} - l_p = \sqrt{\left[l_z.\sin(\alpha_z) + \delta_p\right]^2 + \left[l_z.\cos(\alpha_z) + \delta_{r0}.\cos(j.\gamma)\right]^2} - l_p \tag{15}$$

$$\cos(\alpha_j) = \frac{l_z.\cos(\alpha_z) + \delta_{r0}.\cos(j.\gamma)}{\sqrt{\left[l_z.\sin(\alpha_z) + \delta_p\right]^2 + \left[l_z.\cos(\alpha_z) + \delta_{r0}.\cos(j.\gamma)\right]^2}} \tag{16}$$

By loading the pre-stressed bearing with a radial force, the distance, O_AO_{ip}, between the centre of the balls is constant, (Figure 8 b, c).

$$l_p.\sin(\alpha_p) = l_{rj}.\sin(\alpha_{rj}) = konst. \tag{17}$$

The dependence between the deformation of the jth ball and the top ball can be determined by the relation

$$\delta_j = \delta_0.\cos(j.\gamma) \tag{18}$$

By derivation of equation (14) we get

$$\frac{d\,F_{r1}}{d\,\delta_0} = i.\sum_{j=0}^{z}\left[\frac{d\,P_j}{d\,\delta_j}.\cos(\alpha_j) - P_j.\sin(\alpha_j).\frac{d\,\alpha_j}{d\,\delta_j}\right].\frac{d\,\alpha_j}{d\,\delta_0}.\cos(j.\gamma) \tag{19}$$

The unknown derivatives in equation (19) can be calculated by changing the relations (11), (17), (18).

$$\frac{dP_j}{d\delta_j} = \frac{3}{2}k_\delta^{2/3}P_j^{1/3} \tag{20}$$

$$\frac{d\alpha_j}{d\delta_j} = -\frac{tg(\alpha_j)}{l_{rj}} \tag{21}$$

$$\frac{d\delta_j}{d\delta_0} = \cos(j.\gamma) \tag{22}$$

The interdependence of the contact deformation and radial displacement, Figure 8, can be determined from the relation

$$\frac{d\,\delta_0}{d\,\delta_{r0}} = \left(\frac{d\,\delta_j}{d\,\delta_0}\right)^{-1}.\frac{d\,\delta_j}{d\,\delta_{r0}} \tag{23}$$

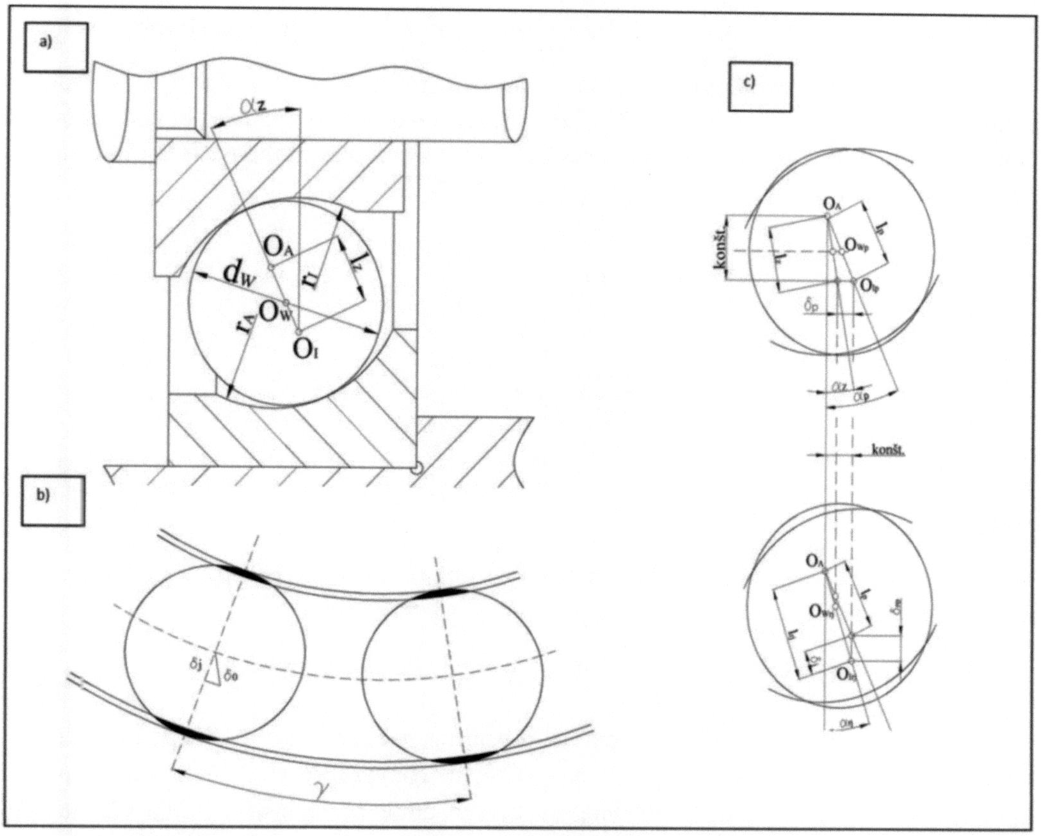

Figure 8. Detailed bearing scheme, a – unloaded, b – pre-stressed, c – radial loaded

Where $\frac{d\,\delta_j}{d\,\delta_{r0}}$ is calculated from equation (15)

$$\frac{d\delta_j}{d\delta_{ro}} = \frac{1}{2}\cdot\frac{2\left(1_z\cos\alpha_z + \delta_{ro}\cos(j.\gamma)\right)\cos(j.\gamma)}{\sqrt{\left(1_z\cos\alpha_z + \delta_{ro}\cos(j.\gamma)\right)^2 + \left(1_z\sin\alpha_z + \delta_p\right)^2}} = \cos\alpha_j\cos(j.\gamma) \tag{24}$$

by inserting equations (24) and (22) into equation (23)

$$\frac{d\,\delta_0}{d\,\delta_{r0}} = \frac{1}{\cos(\mathrm{j}.\gamma)}.\cos(\alpha_j).\cos(j.\gamma) = \cos(\alpha_j) \tag{25}$$

After inserting equations (25) and (19) into equation (13) we get the resulting relation for the stiffness of a pre-stressed nodal point with directionally-arranged bearings.

$$K_r = i.\sum_{j=0}^{z}\left[\frac{3}{2}.k_\delta^{2/3}.P_j^{1/3}.\cos^2(\alpha_j) + P_j.\frac{\sin^2(\alpha_j)}{l_{rj}}\right].\cos^2(j.\gamma) \tag{26}$$

3.3. Stiffness of nodal point with bearings arranged according to shape

When calculating the nodal point with bearings arranged according to shape, we divide the nodal point into part "1" and part "2" (Table 1), with the *same* orientation of contact angles in nodes with directionally-arranged bearings, and the stiffness of the parts is calculated as follows:

$$K_{r1}=i_1.\sum_{j=0}^{z}\left[\frac{3}{2}.k_{\delta}^{2/3}.P_j^{1/3}.\cos^2\left(\alpha_{1j}\right)+P_j.\frac{\sin^2\left(\alpha_{1j}\right)}{l_{r1j}}\right].\cos^2\left(j.\gamma\right) \quad \text{(a)}$$

$$K_{r2}=i_2.\sum_{j=0}^{z}\left[\frac{3}{2}.k_{\delta}^{2/3}.P_j^{1/3}.\cos^2\left(\alpha_{2j}\right)+P_j.\frac{\sin^2\left(\alpha_{2j}\right)}{l_{r2j}}\right].\cos^2\left(j.\gamma\right) \quad \text{(b)} \tag{27}$$

For example in Figure 9 the total numbers of bearings in the front node SBS is 5: $i_1 = 3$, $i_2 = 2$, *contact angles* $\alpha_1 = \alpha_2 = 25$.

We determine the total stiffness of the nodal point by the addition of both parts of the node with the equation:

$$K_r = K_{r1} + K_{r2} \tag{28}$$

In order to optimize the stiffness and load-bearing capacity for specified technological conditions, the manufacturers of machine tools have come out with a new, non-traditional solution for nodal points. By diminishing the contact angle of the bearing in Part 2, the axial stiffness of the nodal point is partially decreased, but at the same time, the value of the radial stiffness and boundary axial load is increased.

3.4. Approximate calculation of stiffness

When evaluating the overall stiffness of a spindle, the designer must take into account the approximate calculation of the stiffness of the nodal points.

If all the balls are loaded, and there are more than 2 per bearing [14], the following equation can be applied:

$$\sum_{j=0}^{z}\cos^2\left(j.\gamma\right)=\frac{z}{2} \tag{29}$$

If the bearing angle is loaded only in an axial direction by the pre-stressing force, then the load on the rollers is constant around the whole circumference and can be expressed, for the particular parts of the nodal point [11] in the form

$$P_{1j}=\frac{F_p}{i_1.z.\sin\left(\alpha_{p1}\right)}; \qquad P_{2j}=\frac{F_p}{i_2.z.\sin\left(\alpha_{p2}\right)} \tag{30}$$

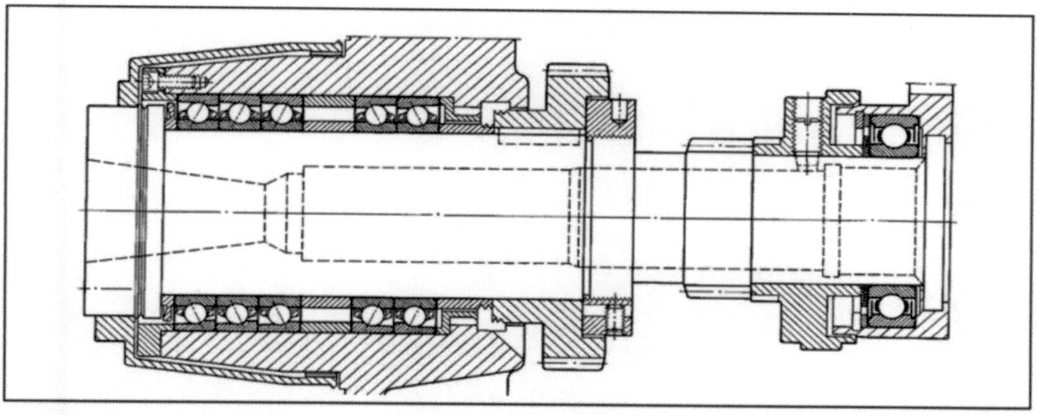

Figure 9. Horizontal machining centre, Thyssen-Hüller Hille GmbH , Germany; Work nodal – 3x71914 ACGB/P4 - 2x71914 ACGB/P4, Opposite side– 6011-2Z

If the magnitude of the spindle bearing contact angles is not greater than 26 degrees, then the value of the second expression in equations (27a) and (27b) is negligible.

Taking these assumptions into consideration, we obtain the relationship for the approximate calculation of the radial stiffness of a bearing angle with directionally placed bearings in the form:

$$K_r = \frac{3.10^{-3}}{4}.z^{2/3}.k_\delta^{2/3}.i^{2/3}.F_p^{1/3}.\frac{\cos^2(\alpha)}{\sin^{1/3}(\alpha)} \tag{31}$$

and with bearings arranged according to shape in the form:

$$K_r = \frac{3.10^{-3}}{4}.z^{2/3}.k_\delta^{2/3}.i_1^{2/3}.F_p^{1/3}.\frac{\cos^2(\alpha_1)}{\sin^{1/3}(\alpha_1)}.\left[1+\frac{i_2^{2/3}.\cos^2(\alpha_2).\sin^{1/3}(\alpha_1)}{i_1^{2/3}.\cos^2(\alpha_1).\sin^{1/3}(\alpha_2)}\right] \tag{32}$$

where the approximate value of the deformation constant is

$$c_\delta = 10^5.\sqrt{1,25\ .\ d_w} \tag{33}$$

d_w – is the diameter of the balls.

The pre-stressing value "F_p" can be calculated according to the standard, STN 02 46 15. Some foreign manufacturers (for example, SKF, FAG, SNFA ...) publish this value in their catalogues. The number of balls "z" and their diameters "d_w" of some types of bearings are quoted in the literature, e.g. [16].

- Based on the equation for the calculation of a nodal point axial stiffness [17]

$$K_a = \frac{3.10^{-3}}{2} z^{\frac{2}{3}} \cdot k_\delta^{\frac{2}{3}} \cdot i_1^{\frac{2}{3}} \cdot F_p^{\frac{1}{3}} \cdot \sin^{\frac{5}{3}}\alpha_1 \left[1+\frac{i_2^{\frac{2}{3}} \cdot \sin^{\frac{5}{3}}\alpha_1}{i_1^{\frac{2}{3}} \cdot \sin^{\frac{5}{3}}\alpha_2}\right] \tag{34}$$

and substituting the equation in brackets

$$T_1 = 1 + \frac{i_2^{\frac{2}{3}} \cos^2 \alpha_2 \sin^{1/3} \alpha_1}{i_1^{\frac{2}{3}} \cos^2 \alpha_1 \sin^{1/3} \alpha_2} \text{ in equation} \tag{35}$$

and

$$T_2 = 1 + \frac{i2^{\frac{2}{3}} \sin^{\frac{5}{3}} \alpha_2}{i1^{\frac{2}{3}} \sin^{\frac{5}{3}} \alpha_1} \text{ in equation} \tag{36}$$

the dependence between the axial and radial stiffness can be expressed by the relation

$$K_r = \frac{K_a}{2} \cdot \frac{1}{tg2\alpha 1} \cdot \frac{T_2}{T_1} \tag{37}$$

When $\alpha_1 = \alpha_2$ in a nodal point with bearings arranged according to shape, or i = 0 in nodal points with bearings arranged according to direction, the quotient of the constants T1, T2 will be equal to 1 and the relation (37) will be simplified. Thus

$$K_r = \frac{K_a}{2} \cdot \frac{1}{tg2\alpha} \tag{38}$$

Taking equations (32) and (34) into consideration, it is evident that the stiffness of the bearing arrangement depends on the number of bearings (i_1 and i_2) in the arrangement, the dimensions of the bearings (z_1, d_{w1} and z_2, d_{w2}), the contact angle (α_1 and α_2) and the preload value F_p.

3.5. Conclusions of the analysis

The conclusions of the analysis [10] are as follows:

- Radial stiffness increases proportionally with increasing values of "z", "d_w", "i", "F_p" and decreases when "a" is prolonged, (Figure 10).
- The parameters "z" and "d_w" must be evaluated in mutual interaction because they characterize the size and dimensions of the bearings. Increasing both of these parameters, and producing the consequent increase of stiffness of the bearing arrangement, can be achieved by increasing the inner bearing diameter. The disadvantage here is that maximum revolving speed will be reduced. A more suitable solution is to decrease the width of the bearing, e.g. from B72 to B70, B719 or B718. In this case the number of rolling elements "z" will be increased and their diameter "d_w" will be smaller.
- Considering equations (31), (32) and (34), it is evident that "z" has a more important influence on stiffness than "d_w". If the diameter of the rolling elements is smaller, their

weight will also be decreased, and this fact will allow an increase in the maximum revolving speed.

- The number of bearings in bearing arrangement "i" is the significant factor which can favourably influence stiffness. But the increased number of bearings will reduce the maximum revolving frequency and therefore it is possible to use this solution only for low speed spindle-bearing systems.
- The preload has a relatively small effect on the stiffness of bearing arrangements. The preload real value also depends on the type of flange used. When fixed flanges are used the preload value can exceed the nominal value by several times. This will cause excessive preload values which produce heat and the bearing arrangement will break down much sooner than expected.
- The contact angle "α" has a significant influence on the variation of the stiffness of the bearing arrangement. When the value of the contact angle is increased, the radial stiffness and maximum revolving speed of the bearing arrangement is also decreased. On the other hand, the axial stiffness of the bearing arrangement will be significantly increased.

4. Optimization of the spindle-bearing system in relation to temperature

In addition to the bearing arrangements, the temperature properties of the bearing supporting node have an increasingly greater significance on the high-speed capability of the bearing. The main goal of this section is to show the SHS design under real operating conditions, taking into consideration the temperature-related behaviour of the spindle and bearing nodes.

The value of the changes in SHS temperature depends on the temperature gradient, the type of bearing arrangement (DB, DF, DT, ...), the contact angle of the bearing, and the distance between the bearings arranged in the node.

The stiffness of the given example was analysed using the application software "*Spindle Headstock*" [3], developed in our department.

The analysis identified the optimal stiffness, which was then applied to the headstock of the DB 24 fy. Ex-Cell-O GmbH., Eislinger precision boring machine, Figure 11, [18]

The headstock used for analysis had the following parameters:

Output power:	$P = 3$ kW
Maximum speed:	$n_{c\,max} = 5500$ min^{-1}
Shape inaccuracy at boring	$E_s \leq 1{,}5$ μm
Surface roughness:	$E_t = 1...1{,}5$ μm
Bearings:	FAG B 7016 C.TPA.P4.UL in "O" arrangement
Bearing lubrication:	grease

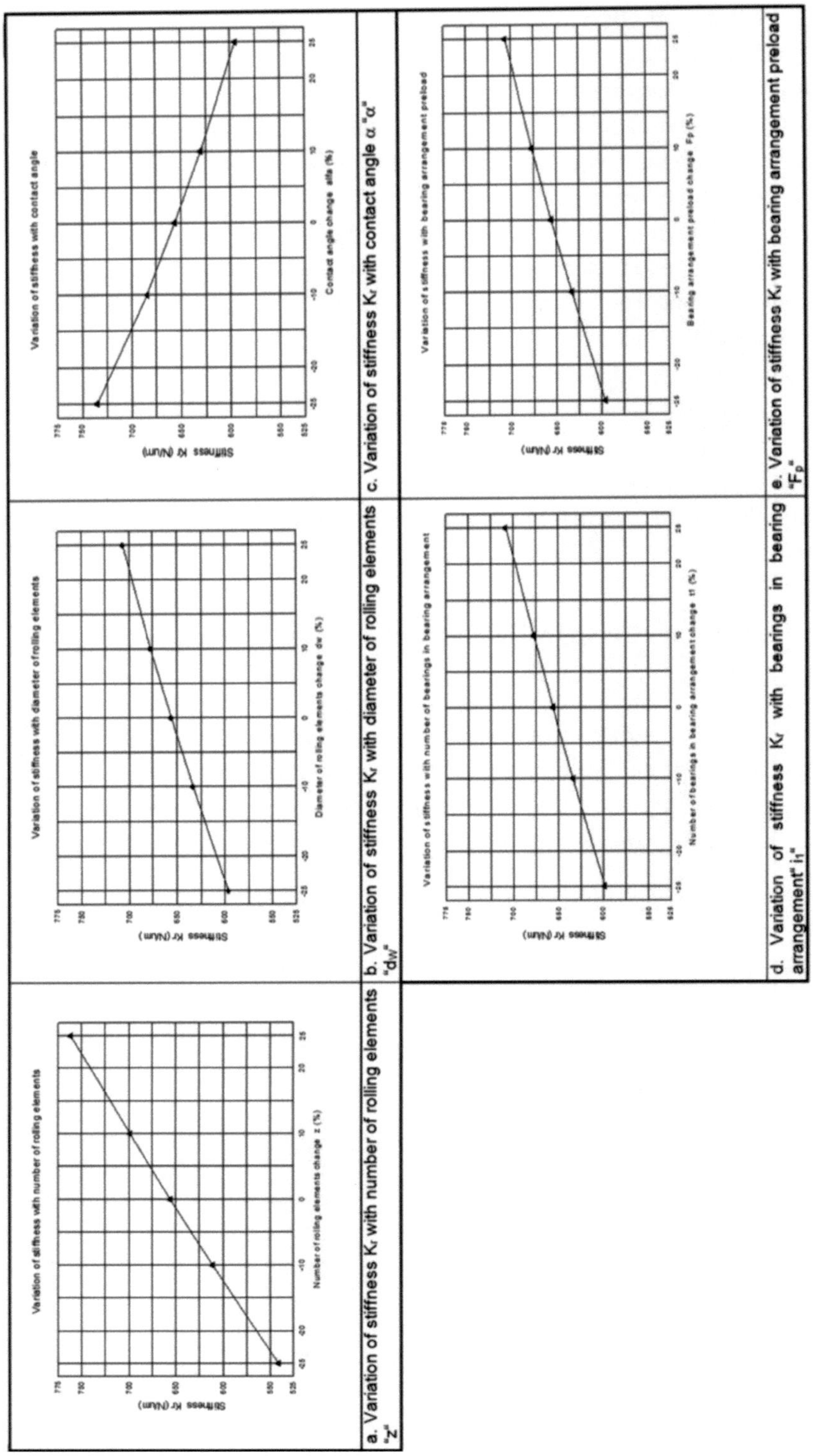

Figure 10. Variation of stiffness with bearing arrangement parameters

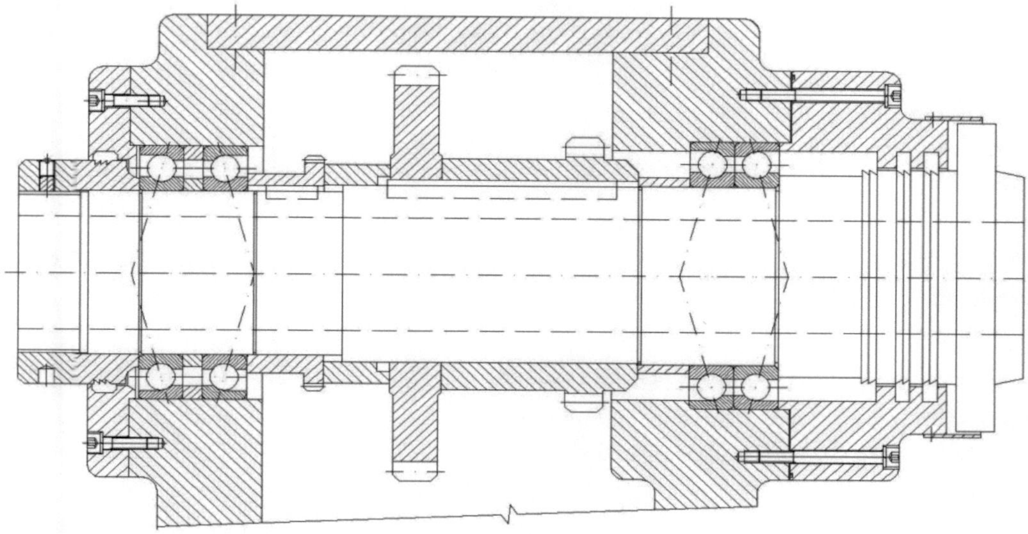

Figure 11. The Headstock of the precision boring machine DB 24 fy. Ex-Cell-O GmbH., Eislinger, [18]

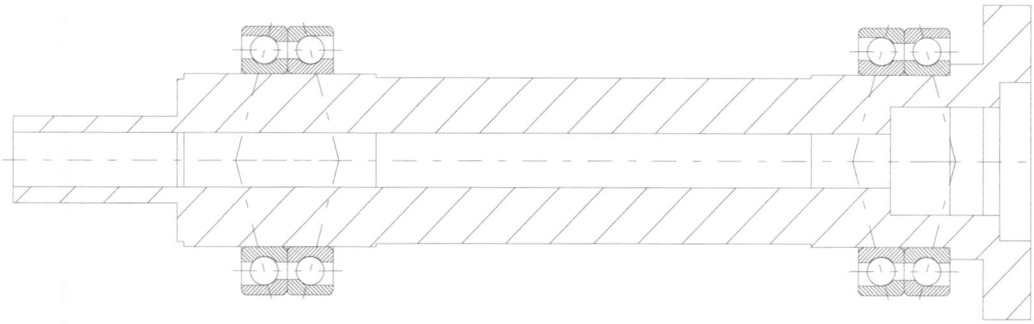

Figure 12. Model of the spindle

Results for spindle with arrangement DB - DB.

Radial load: $F_r = 1000$ N
Axial load: $F_a = 200$ N
Desired spindle speed: $n_{Ot} = 5500$ min^{-1}

No driving force

Working conditions
Lubrication: Plastic grease
Cooling: Good cooling

	Rear support	Front support
Bearings		
- type.	2pc. [B 7016 CTB]	2pc. [B 7016 CTB]
- dimensions [mm]:	D= 125 d= 80 B= 24 dW=13.49	D=125, d=80, B=24, dw=13.49
- arrangement:	<>	<>
- precision grade:	P4	P4
Preload:	Light	Light
Flange:	Fixed flange	Fixed flange
Maximum speed:	$Zn_{max} = 5\,256\ \text{min}^{-1}$	$Zn_{max} = 5\,256\ \text{min}^{-1}$
Pre-load:	$ZF_p = 404$ N	$PF_p = 402$ N
Reactions:	$R_A = 205$ N	$R_B = 1\,205$ N
Radial stiffness:	$K_{rA} = 666\,243\ \text{Nmm}^{-1}$	$K_{rB} = 651\,216\ \text{Nmm}^{-1}$
Axial stiffness:	$K_{aA} = 97\,860\ \text{Nmm}^{-1}$	$K_{aA} = 97\,745\ \text{Nmm}^{-1}$
Durability:	$T_{rvZ} = 394\,366$ hours	$T_{rvP} = 225\,577$ hours
Bearings distance:	$L = 297$ mm	
Total displacement at the end:	$y_{r(L+a)} = 0.00372931$ mm	
Total stiffness:	$\mathbf{K_{rc} = 268\,146\ Nmm^{-1}}$	
Optimal calculated values		
Optimal bearing length:	$L_{opt} = 283.6$ mm	
Optimal displacement at the end L_{Opt}:	$y_{rmin} = 0.00372686$ mm	
Optimal stiffness:	$\mathbf{K_{rcopt} = 268\,322\ Nmm^{-1}}$	

4.1. The optimisation of SHS with regard to temperature

The temperature dilatation of the spindle can be described by the equation:

$$\Delta L = \lambda_t . L . \Delta t \tag{39}$$

If the distance between the bearings in the "DB" arrangement is short (Figure 13a), the dilatations in a radial direction is greater, [18]. The temperature gradient causes the dilatation of the inner bearing rings to be greater than that of the outer rings. Consequently, the original preload increase in temperature will be higher in the bearing node. The elevated temperature will influence the temperature gradient, and the preload value could cause bearing node failure.

The preload change was defined by the change in the distance between the centres of the radii of the rolling raceways:

$$l_0 = r_A + r_I - d_w \tag{40}$$

The distance l_t at given temperature gradient in accordance with Figure 13 is

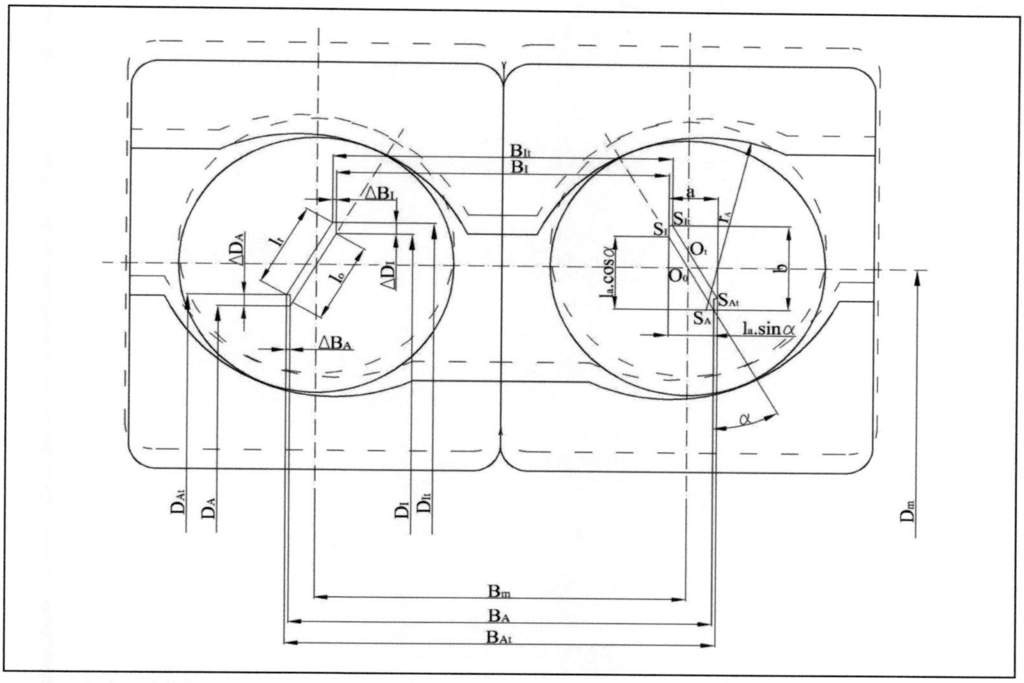

a) Distance between bearings B_m = 29 mm

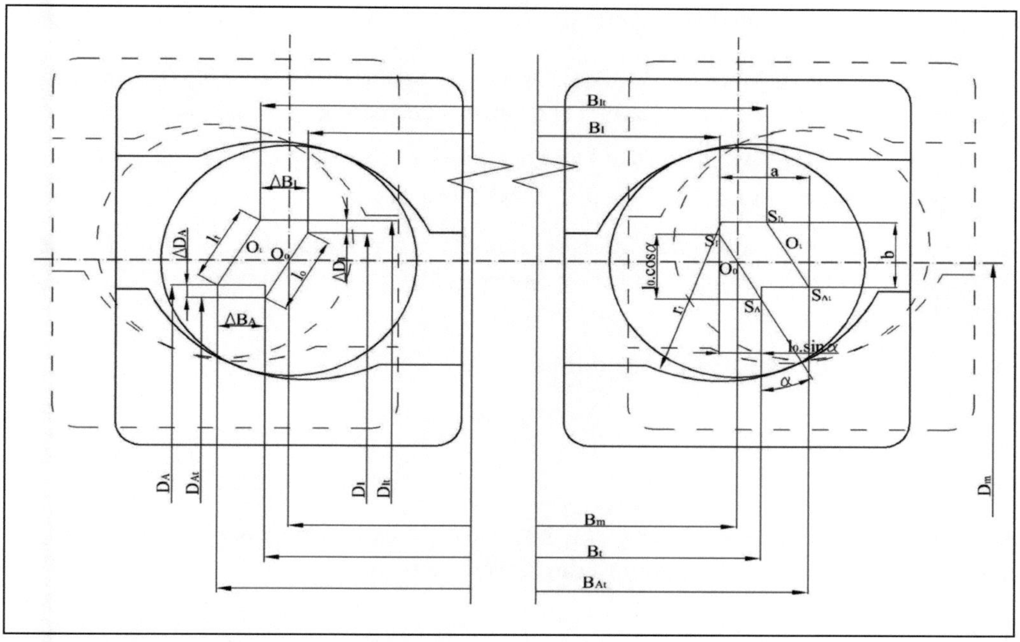

b) Distance between bearings B_m = 500 mm

Figure 13. Temperature deformation of bearing arrangement B 7016 C TPAP4UL in "DB"

$$l_t = \sqrt{a^2 + b^2} \tag{41}$$

where

$$a = l_0.\sin\alpha + \frac{\lambda_t}{2}.\left[B_m.(t_I - t_A) + l_0.\sin\alpha\ .(t_I + t_A - 2.t_0)\right] \tag{42}$$

$$b = l_0.\cos\alpha + \frac{\lambda_t}{2}.\left[D_m.(t_A - t_I) + l_0.\cos\alpha\ .(t_I + t_A - 2.t_0)\right] \tag{43}$$

The magnitude of deformation will be

$$\Delta\delta\ = l_0 - l_t \tag{44}$$

and preload change in accordance with [18] will be

$$\Delta F = \Delta\delta\ .z^{2/3}.c_\delta^{\ 2/3}.\sin^{8/3}\alpha \tag{45}$$

where

$$c_\delta\ = 10^5.\sqrt{1{,}25\ .\ d_w} \tag{33}$$

In the twin bearings FAG B 7016 C.TPA.P4.UL with "DB" arrangement, at a temperature gradient of 10 °C, and with bearing distance B_m = 29 mm, the preload will increase by 13,32 N.

Conversely, if the distance of the bearing in "O" arrangement is long (Figure 13b), the dilatations in the axial direction prevail and cause a decrease in the value of the preload.

In the twin bearings FAG B 7016 C.TPA.P4.UL with "DB" arrangement, at a temperature gradient of 10°C , the preload will be decreased by 5,88 N.

In "DB" arrangement, the main goal of temperature optimization is dependant on the determination of the optimum distance between the bearings at which a change in the preload at the given temperature gradient would be minimal.

In accordance with Figure 13 the condition

$$l_0 = l_t \tag{46}$$

must be satisfied.

By substituting equations (42) and (43) into (46), the optimal bearing separation distance from the point of view of temperature can be deduced from:

$$B_{mopt} = D_m\ .\frac{\cos\alpha}{\sin\alpha}\ -\ \frac{l_0.(t_I + t_A - 2.t_0)}{t_I - t_A}\ .\left(\frac{1}{\sin\alpha}\right) \tag{47}$$

Figure 14 shows the change of optimal bearing distance at various values of the temperature gradient for the analysed SBS, Figure 11.

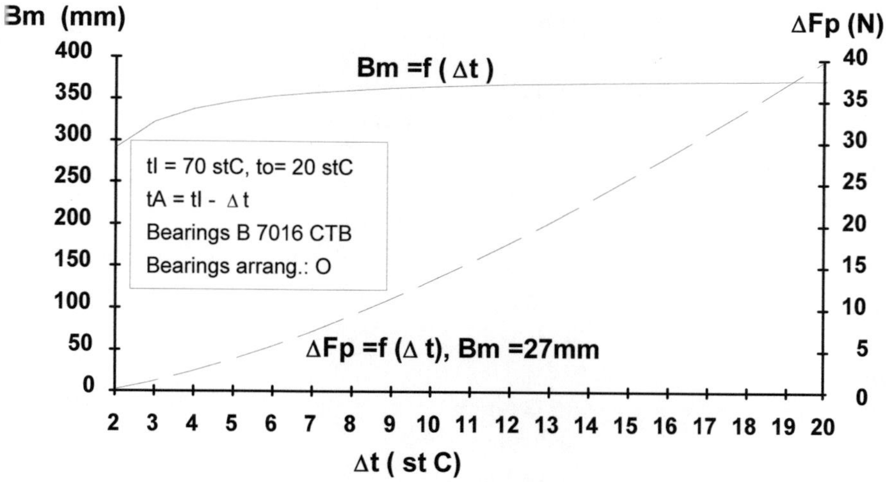

Figure 14. The inter - dependence of bearing preload change, ideal distance between bearings and change of temperature in the bearings arrangement system.

4.2. Recommendation for improvements in construction

The recommendations from the point of view of temperature optimisation for the DB 24 SHS boring machine are based on the results of the analysis undertaken. From the perspective of temperature, it can be seen that a change in bearing node arrangement to individual spindle supports from "DB" to "DT" would be advantageous, Figure 15.

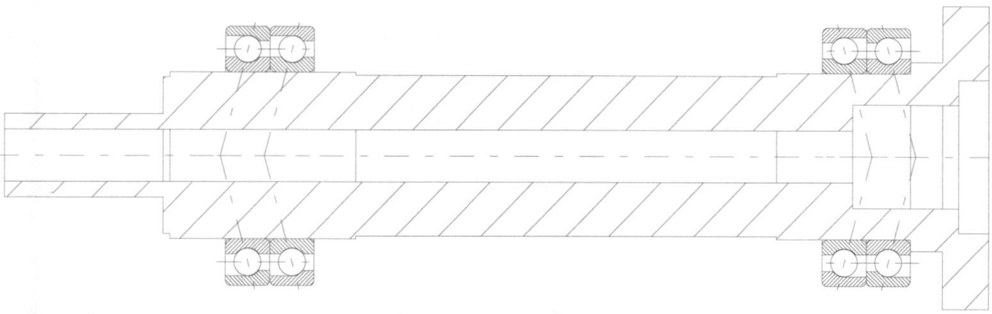

Figure 15. Model of the spindle

Results for spindle with arrangement DB - DB.

Radial load: F_r=1000 N
Axial load: F_a = 200 N
Desired spindle speed: n_{Ot} = 5500 min^{-1}
No driving force

Working conditions

Lubrication: Plastic grease
Cooling: Good cooling

	Rear support	**Front support**
Bearings		
- type.	2pc. [B 7016 CTB]	2pc. [B 7016 CTB]
- dimensions [mm]:	D= 125 d= 80 B= 24 dW=13.49	D=125, d=80, B=24, dw=13.49
- arrangement:	<>	<>
- precision grade:	P4	P4
Preload:	Light	Light
Flange:	Fixed flange	Fixed flange
Maximum speed:	$Zn_{max} = 5\,256$ min^{-1}	$Zn_{max} = 5\,256$ min^{-1}
Pre-load:	$ZF_P = 404$ N	$PF_P = 402$ N
Reactions:	$R_A = 205$ N	$R_B = 1\,205$ N
Radial stiffness:	$K_{rA} = 666\,243$ Nmm^{-1}	$K_{rB} = 651\,216$ Nmm^{-1}
Axial stiffness:	$K_{aA} = 97\,860$ Nmm^{-1}	$K_{aA} = 97\,745$ Nmm^{-1}
Durability:	$T_{rvZ} = 394\,366$ hours	$T_{rvP} = 225\,577$ hours

Bearings distance:	$L = 297$ mm
Total displacement at the end:	$y_{r(L+a)} = 0.00372931$ mm
Total stiffness:	$\mathbf{K_{rc} = 268\,146}$ **Nmm^{-1}**

Optimal calculated values

Optimal bearing length:	$L_{opt} = 283.6$ mm
Optimal displacement at the end L_{Opt}:	$y_{rmin} = 0.00372686$ mm
Optimal stiffness:	$K_{rcopt} = 268\,322$ Nmm^{-1}

In comparison with the original bearing node arrangement, the radial stiffness of the rearranged spindle-bearing system will drop slightly, but its axial stiffness will increase. The advantage of the reconfigured SBS is that at real mean values of temperature gradient, the SBS stiffness will be almost fixed.

4.3. Spindle headstock

The application software is used for calculating the SBS of machine tools supported on rolling bearings. The programme enables us to determine all elements and calculate the properties of the spindles and shafts which are supported on rolling bearings. The application software enables very fast and user-friendly calculation of the radial spindle stiffness in the bearing arrangement in a bearing unit.

The architecture of the programme contains a number of mathematical formulae which have been experimentally verified. These models respect the conditions of the spindle working accuracy in terms of the external load cutting forces, driving forces, and also spindle rotation speed.

The basic interactive programme offers:

1. The ability to input user-determined conditions for the calculation and optimisation of the spindle fitting system (Figure 16);

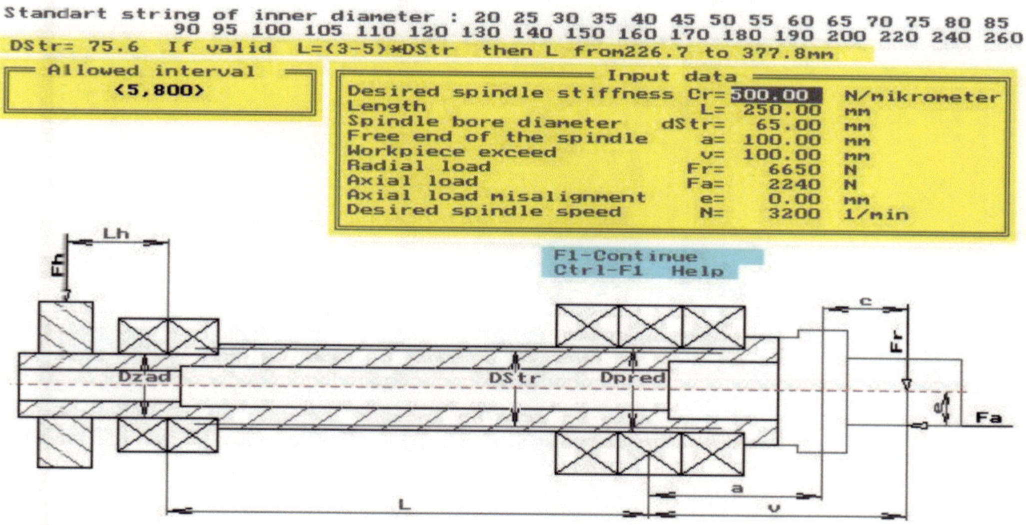

Figure 16. Entering the input data

2. The ability to select the most appropriate bearing or bearing node arrangements (Figure 17, Figure 18, Figure 19). Data about selected bearings can be ganed from extensive databases according to the users requirements within the bearing inner diameter range:

- angular contact ball bearings, type 7 10..150 mm
- single row cylindrical roller bearings, type N 50..120 mm
- full cylindrical roller bearings, type NN 30..440 mm
- axial angular contact bar, type 2344 25..380 mm
- thrust ball bearings, single direction, type 51 10..360 mm
- thrust ball bearings, double direction, type 52 10..190 mm
- deep groove ball bearings, type 6 3..360 mm;

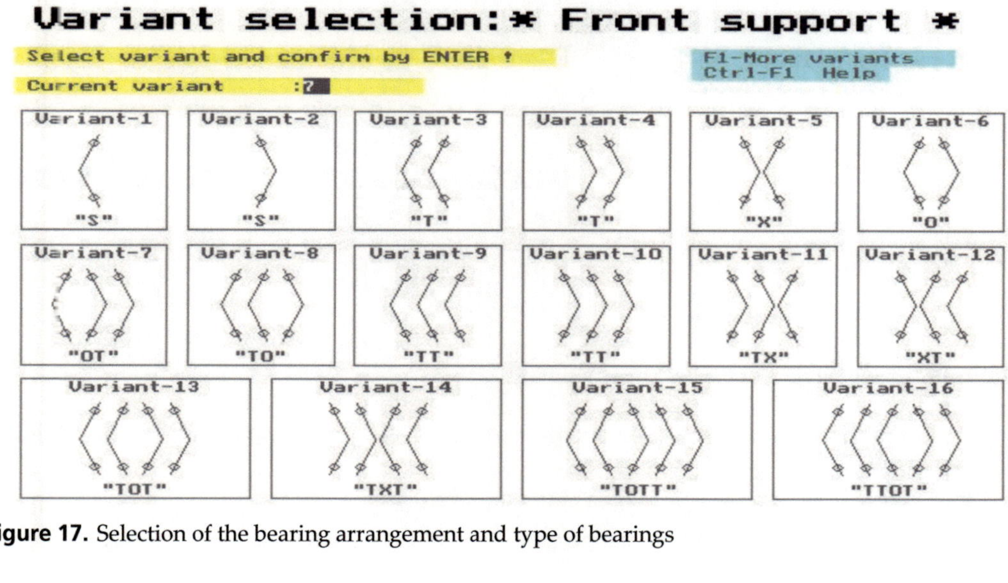

Figure 17. Selection of the bearing arrangement and type of bearings

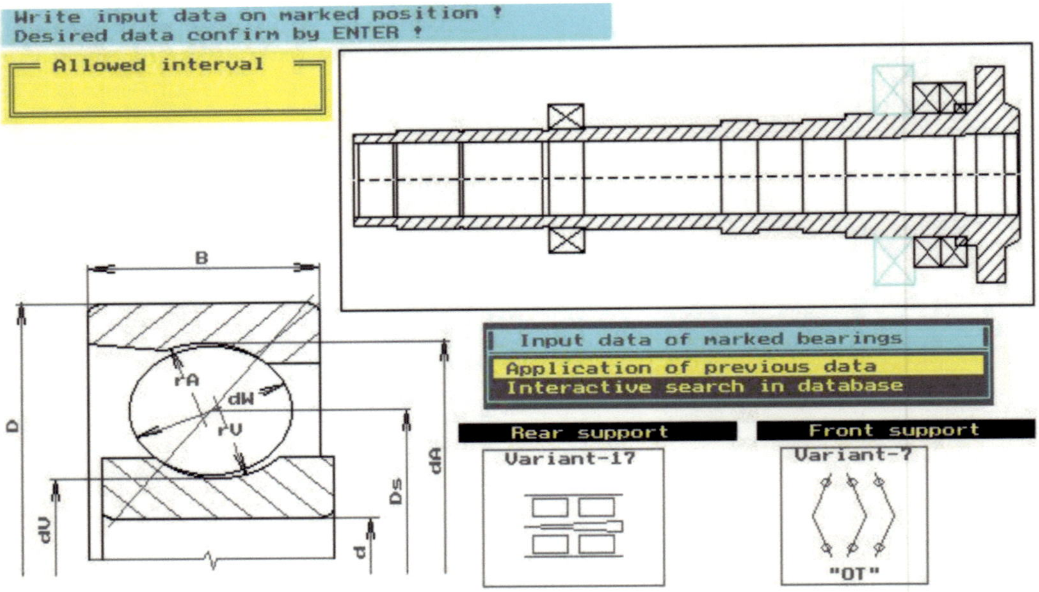

Figure 18. Dimensional design data of the spindle housing

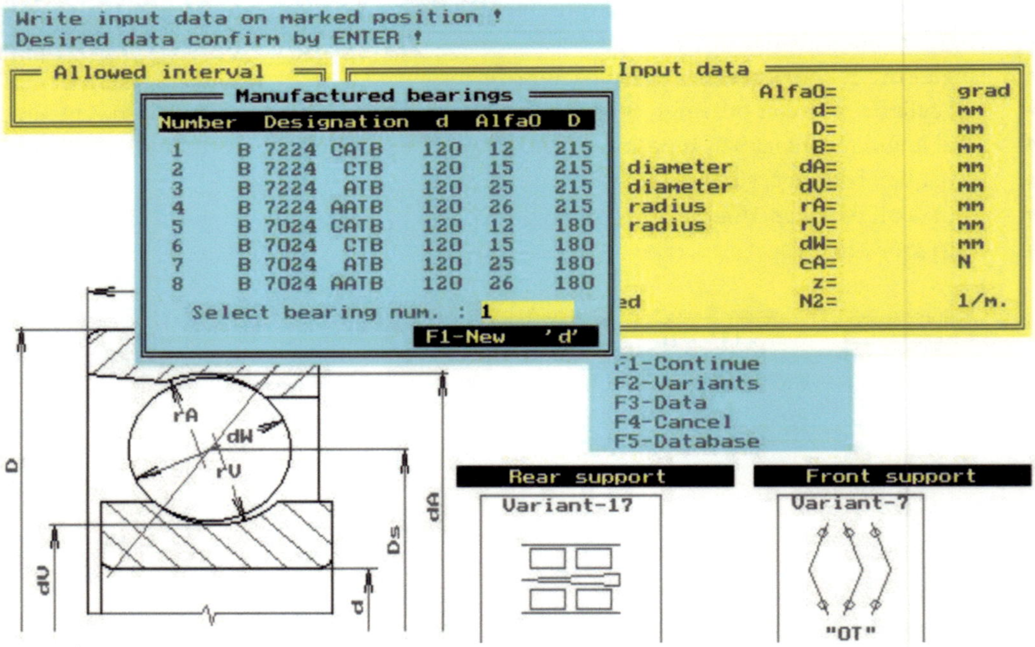

Figure 19. Changing data of the bearings mounting

3. The identification and selection of the standardized spindle nose for turning, milling, grinding and boring;
4. The choice of the design parameters and spindle suitability for different working conditions (working accuracy, preloading, flange type, lubrication system, cooling), Figure 20;

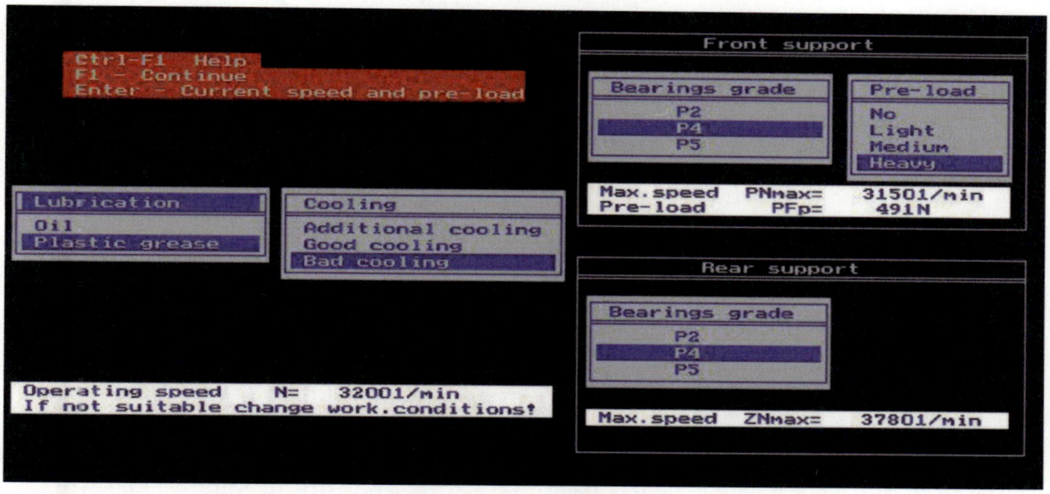

Figure 20. Entering preliminary data for the bearings conditions

5. The calculation and optimization of the cutting parameters for the required material to be machined (cutting force, torque, feed , power), Figure 21;
6. Calculation and optimization of the design and fitment with regard to the applied conditions (revolving speed, radial stiffness, axial stiffness, rating life) for the bearing units and the fitting as a whole, for all of the identified bearing types.

Graphical output of partial deflections caused by bearing nodes distance are shown in Figure 21.

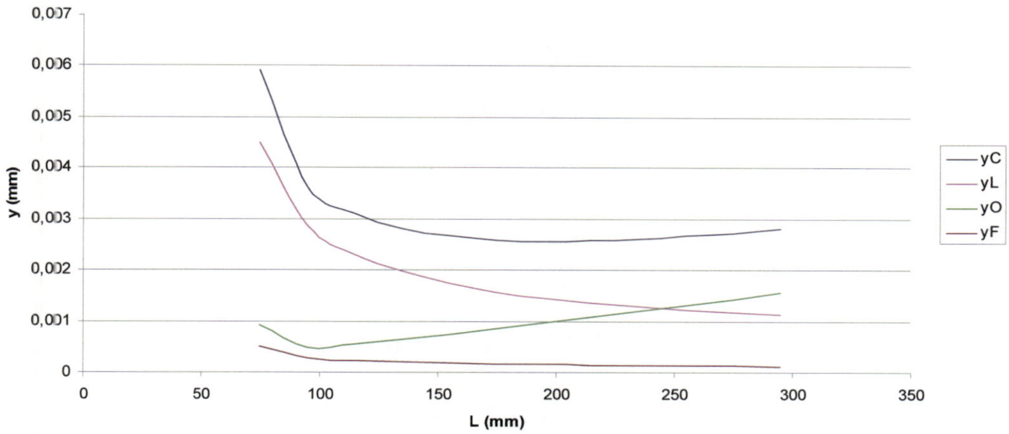

Figure 21. Graphical output of the dependence of partial deflection on bearing node distance

The results include:

- applied entries
- chosen bearing (unit) fit types
- rotational speed limit
- radial unit stiffness

- axial unit stiffness
- resultant stiffness for the chosen spindle fit system
- load parameter and durability
- graphical illustration of the chosen design (Table 1.).

The "Spindle Headstock" software application product is a basic programme which can be modified according to the user's wishes.

The programme has been written in the source code programming language, T-PASCAL v.7, with special additional modules for graphics. A number of interactive modules prepare user-specified data for use in AutoCAD utilising the DXF format. The programme can be used on any IBM/PC compatible computer using a HERCULES, EGA or VGA graphics adapter.

The applied software technology has been used in the industry to improve the working accuracy of the machine tools made by TOS Trenčín-Slovakia, for SN and SPSI type lathes , (2), and to design the boring headstocks for the modular single-purpose machine tools made by TOS Kuřim-Czechoslovakia, (5) TOS Lipník, SKF, GMN and INA Skalica. The programme is very effective and reliable and comparison of the results between experiments and calculations show good correlation, never exceeding 10 %.

5. Experimental research

Theoretical results and hypotheses must be verified by experimental tests.

5.1. Research of bearing nodes characteristics

In some cases it is very difficult, or even impossible, to gain experimental results from actual machine tools. This led us to develop an experimental device for research into spindle bearing node arrangement characteristics. Our department has developed such a device that can measure:

1. changes in the bearing contact angle at its mounting point, changes in loading, and changes in revolving frequency,
2. deformation from axial and radial loading for various preload arrangements, contact angles and bearing node revolving speed settings
3. increases in the temperature in the bearing nodes at various settings
4. dimension of cutting forces in bearing nodes

The variation in the stiffness of the bearing arrangement B7216 is shown in Figure 22, [19]. We can use the experimental measuring head for measuring the deflection and temperature of a varying number of bearings and bearing nodes (from 2 to 5), their preload value, dimensions, and the contact angle of the bearings with different radial and axial forces used, [19], [20].

We use this device to measure the deformation characteristics of the bearing node with different combinations of bearing arrangement, pre-stressed values, contact angles, loads and revolution frequencies. We use this experimental measuring head for verifying the theoretical calculation and real performances of the bearing node (stiffness, precision running and temperature).

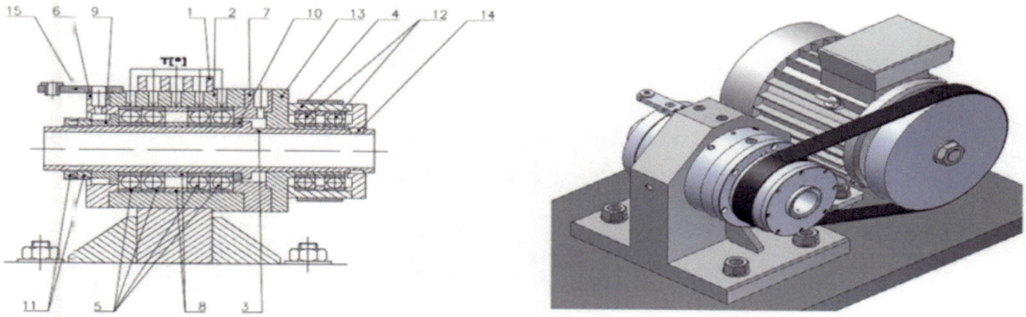

a) cross section of the experimental measuring head with driver, 1- head, 5-bearing node, 4-band wheel

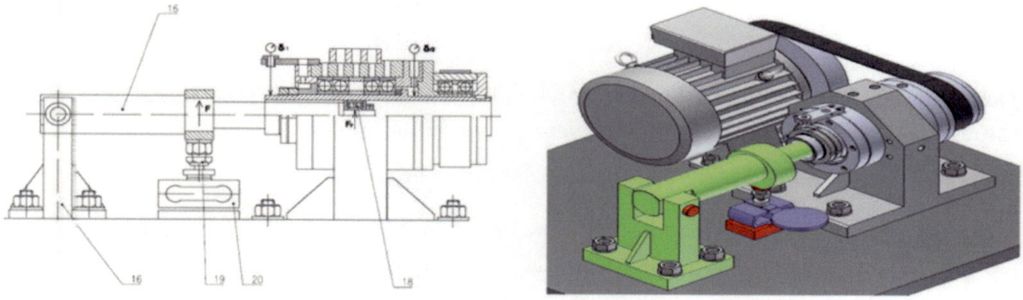

b) arrangement of experimental measuring head for measuring radial stiffness by spindle speed 16 - holder, 19 - tightening screw, 20 - dynamometer, 18 - force bearing

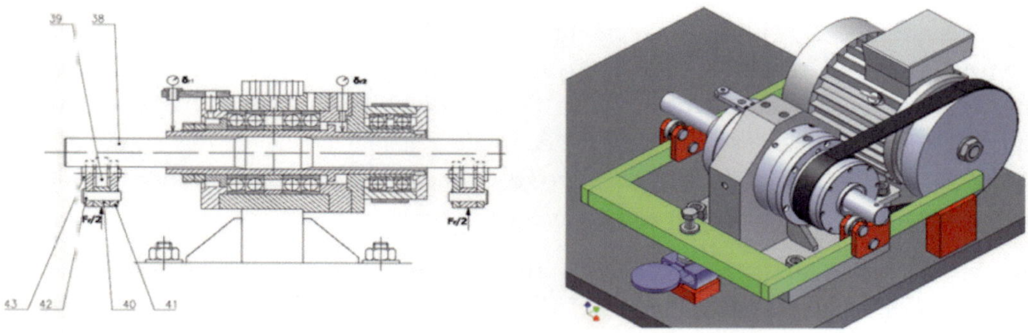

c) arrangement of experimental measuring head for measuring precision running and temperature by spindle speed, 43 - force bearing

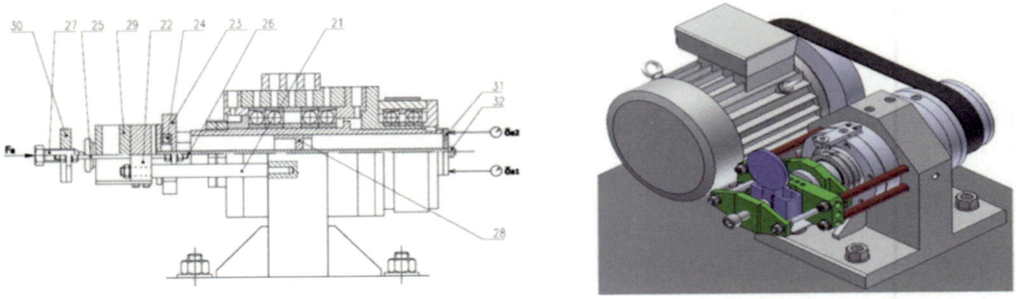

d) arrangement of experimental measuring head for measuring radial stiffness by spindle speed with, 22 - holder, 27 - tightening screw, 29 - dynamometer, 24 - force bearing, 32 - flange

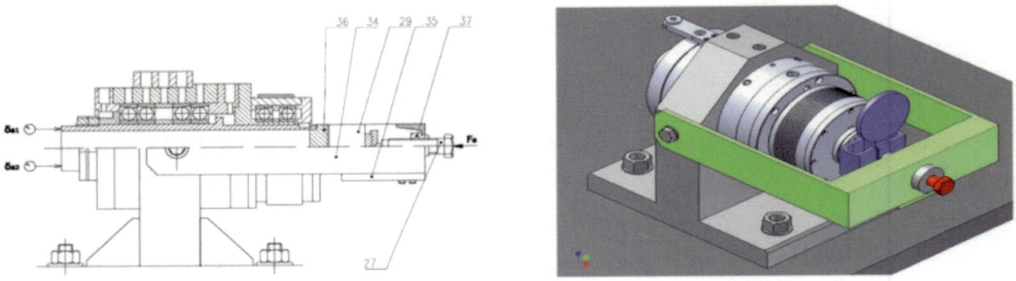

e) arrangement of experimental measuring head for measuring static axial deflection and temperature, 34 - holder, 27 - tightening screw, 29 - dynamometer

Figure 22. Different variants of the experimental measuring head

In Figure 23 we have compared the experimental stiffness measurement, with the accurate theoretical and simplified average calculated radial stiffness of the B7216 AATBP4OUL bearing arrangement. The stiffness variation was examined with a 25% contact angle with nominal value of bearing arrangements: z_1, z_2= 14, d_{w1}, d_{w2} = 19,05 mm, α_1, α_2=12º, F_P=340 N.

When static, the experimental values of radial stiffness are higher than the theoretical values. The dependence of stiffness on loading exhibits a decreasing pattern. The decrease is nearly linear, until a certain critical force "F_{kr}" is reached, at which point the roller with the lightest load becomes unloaded. The deformation characteristic of the nodal point is influenced by the type of flange. The degree and gradation of the stiffness change under the given operational conditions depend on their construction.

In this field the results of the precise and the approximate mathematical model are practically the same. Consequently it follows that in a preliminary mounting design, a simplified mathematical model for calculating the stiffness of the nodal points can be used, as suggested in this article. The convergence of the measured and calculated values provides good evidence for a wider application of the programme in practice.

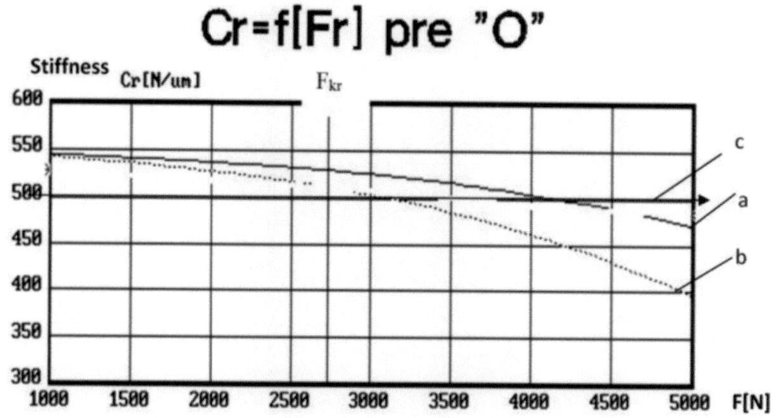

Figure 23. Radial stiffness of the bearing arrangement B7216 AATB P4 O UL, a - experimental, b - accurate theoretical c- simplified average

5.2. New design of headstock

In the new design of a headstock which connects to a CNC system, the maximum width of cut is limited by the point at which self-exciting vibration starts.

From a constructional point of view, the headstock design can be classified as follows:

- classical headstock
- headstock with an integrated drive unit

The classical headstock is a mechanical unit, where a spindle is driven by a motor through a gearbox without any control system.

The disadvantages of the classical construction are as follows:

- problems with the gears at higher revolving frequencies,
- actual cutting speeds are not continual because of the discontinuous nature of the gearboxes,
- large dimensions of complete units

New design "Duplo–Headstock“

The "Duplo-headstock“ has been designed in order to achieve technological parameters comparable to the performance of standard electro-spindles, but at a lower production costs and with higher controllability. This particular headstock is assembled from readily available elements (bearings, single drives,). The demands on the other peripheral devices are reduced, as are the costs.

The “Duplo-headstock” can be described as a spindle with double supports, driven by two separate motors which can operate independently or together. Figure 24 - 28 show a „Duplo–headstock“ design [20].

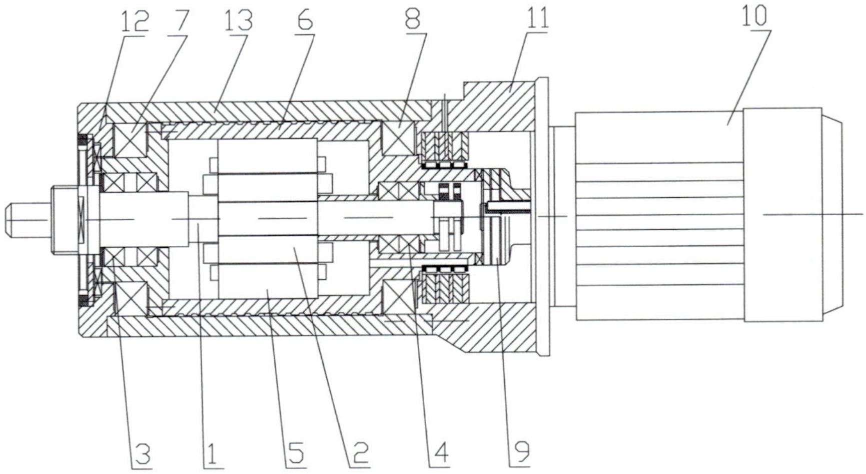

Figure 24. High-speed headstock "Duplo"

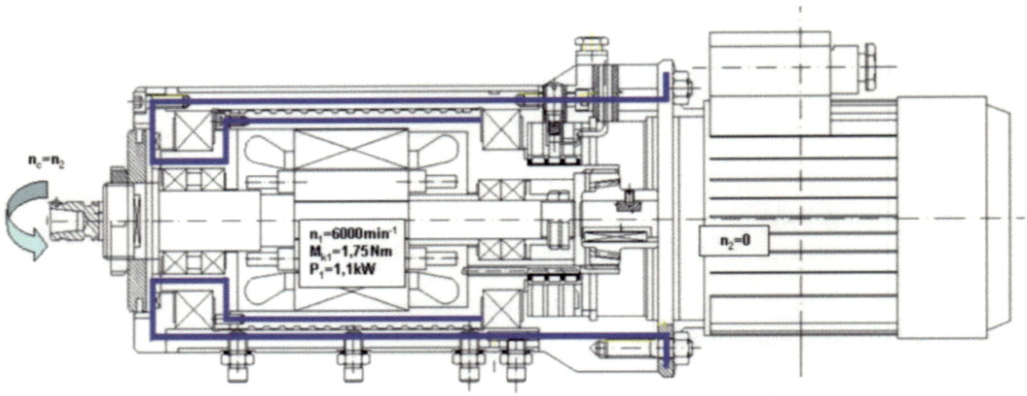

Figure 25. The stator engaged on spindle, Speed: n_{1max}=6000 (min-1); n_2=0, n_{c_spi}=n_1 Torque moment: Mk_{c_sp}=Mk_1=1,75 (Nm) by n_{1max}, Power: P_{c_sp}=P_1=1,1 (kW)

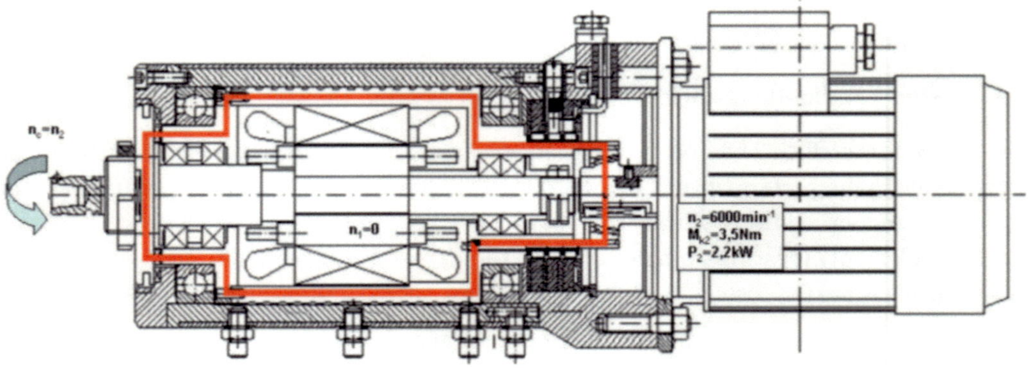

Figure 26. The stator engaged on body, Speed: n_{1max}=0; n_2=6000 (min-1), n_{c_sp}=n_2, Torque moment: Mk_{c_sp}=Mk_2=3,5 (Nm) by n_{2max}, Power: P_{c_sp}=P_2=3,5 (kW)

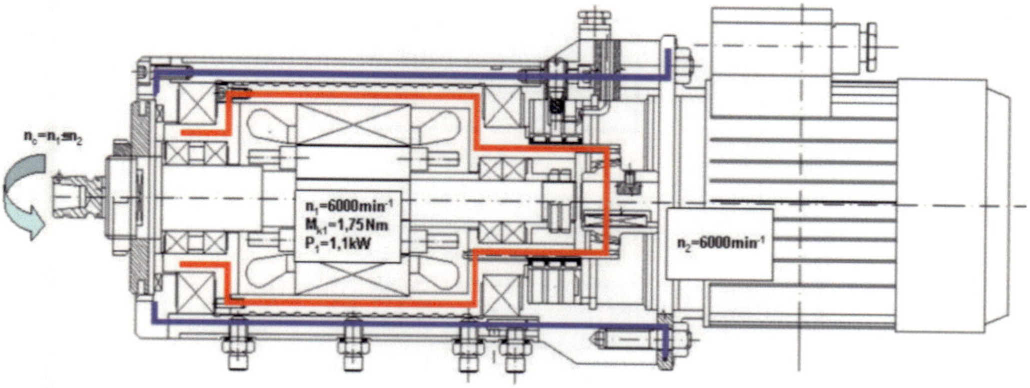

Figure 27. Disengaged, Speed: n_{1max}=6000 (min-1); n_2=6000 (min-1), n_{c_sp}= $n_1 + n_2$ by one direction of rotation, n_{c_sp}= $n_1 - n_2$ by opposite direction of rotation, Torque moment: Mk_{c_sp}=Mk_1=1,75 (Nm) by n_{1max}, Power: P_{c_sp}=P_1=1,1 (kW)

Figure 28. Stand of "Duplo" Headstock

The spindle (1), with built-in armature (2), is supported by bearings (3), (4). The stator of the internal motor (5) is supported on bearings (7), (8). The clutch (9) connects a hollow shaft with an external electro-motor (10). The stator feeding rings (11) are located in the rear part of the shaft. The clutch (12) enabling switching between working modes is located in the front part of the shaft. The advantage of this innovative design, which is already in use, is that the headstock can work in three different modes:

- stator is engaged on the spindle
- stator is engaged on the body
- no engagement

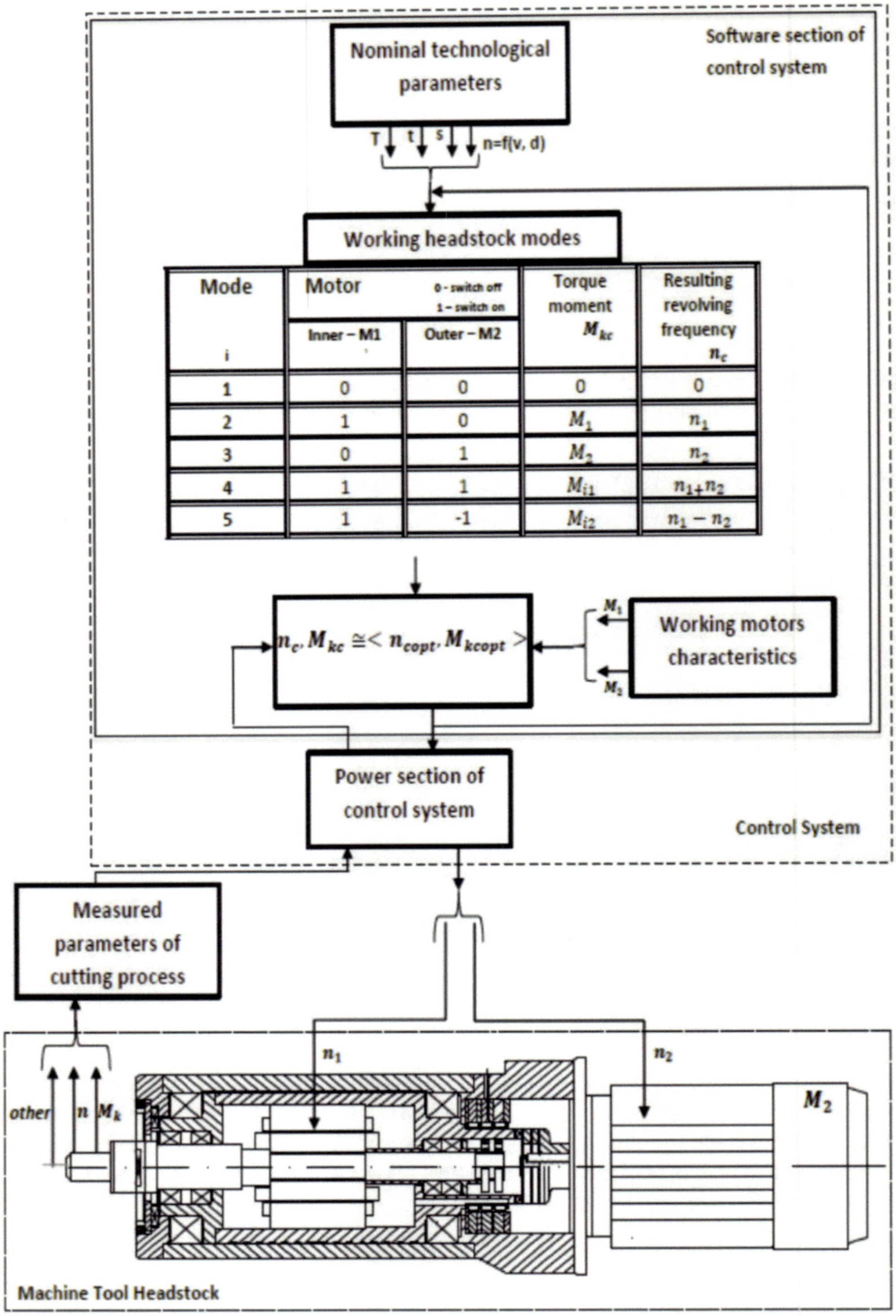

Figure 29. Scheme of Machine Tool Headstock Control

Connecting such a headstock with a suitable control system can provide optimal cutting conditions for various technological operations. The intelligent control system, Figure 29, can operate in any one of the working modes and ensure nominal or optimal technological parameters best suited to the machining process, [21]. Figure 30 shows the design for the construction of the "Duplo" Headstock.

In the third mode, the clutch (12) is switched off. The spindle (1) is driven by both motors, (Figure. 27), providing the maximum speed, which is required, for example, in grinding.

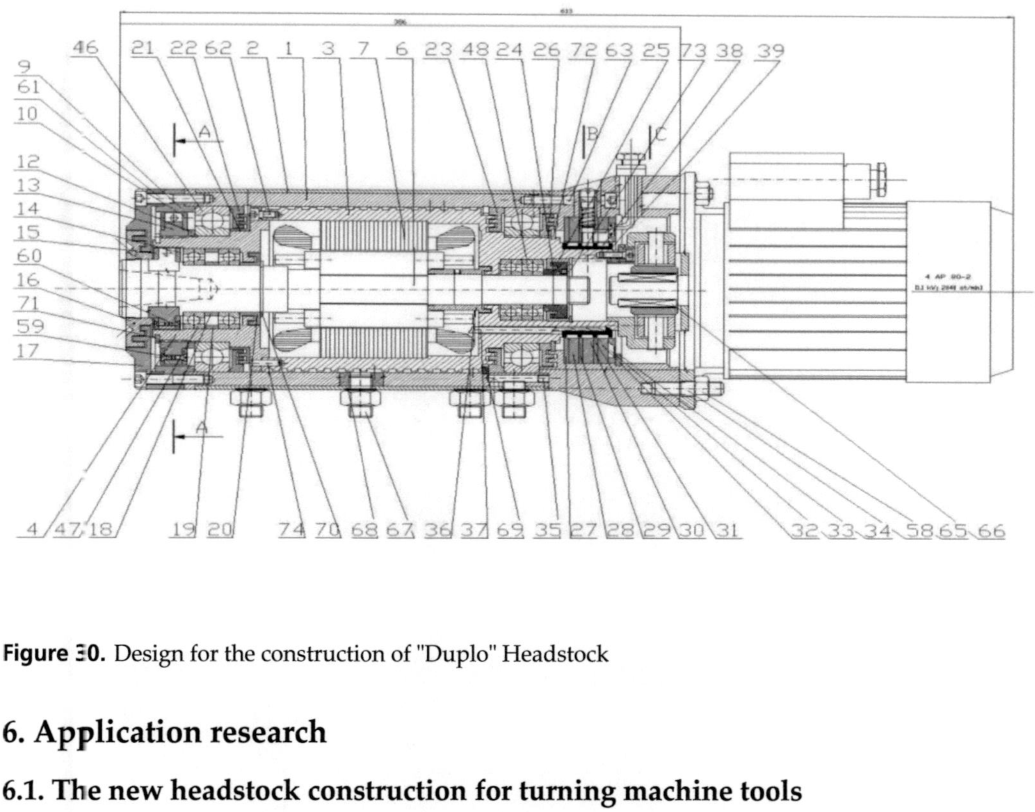

Figure 30. Design for the construction of "Duplo" Headstock

6. Application research

6.1. The new headstock construction for turning machine tools

TRENS a. s. Trenčín, is a Slovak manufacturer of machine tools (mainly lathes) and offers a new generation of lathes implementing various technological advances in design, production, and control systems, [22]. The Department of Production Engineering has been asked to design an accurate running spindle for the SBL 500 CNC lathe (Figures 31 -33), [23]. All construction data and results of measurements were obtained from the producer. Table 2. shows the calculated (Spindle Headstock Version 2.8) values [23] of the optimized design. Figure 34 shows the comparison between the original and optimized designs.

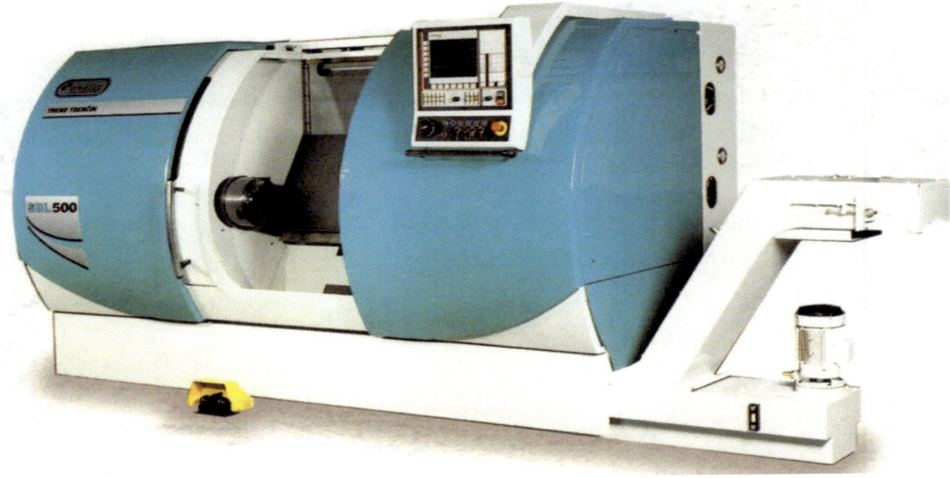

Figure 31. CNC Lathe SBL 500

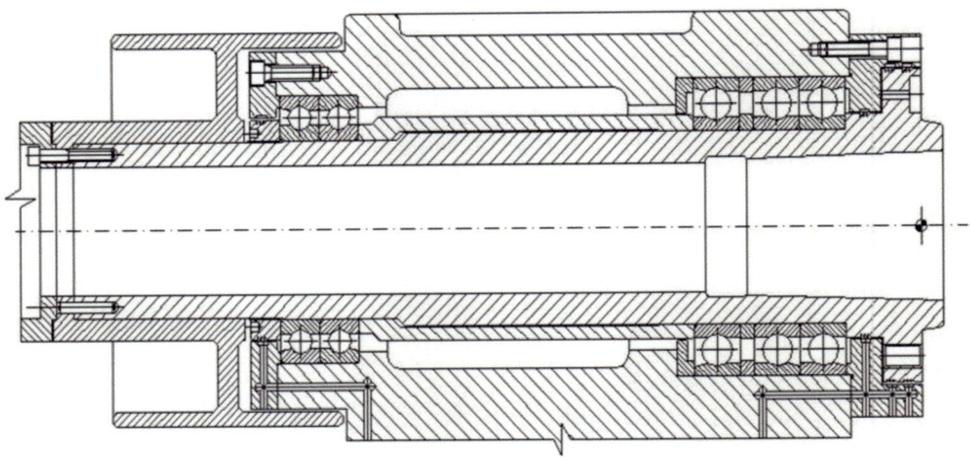

Figure 32. Original design of SBL 500

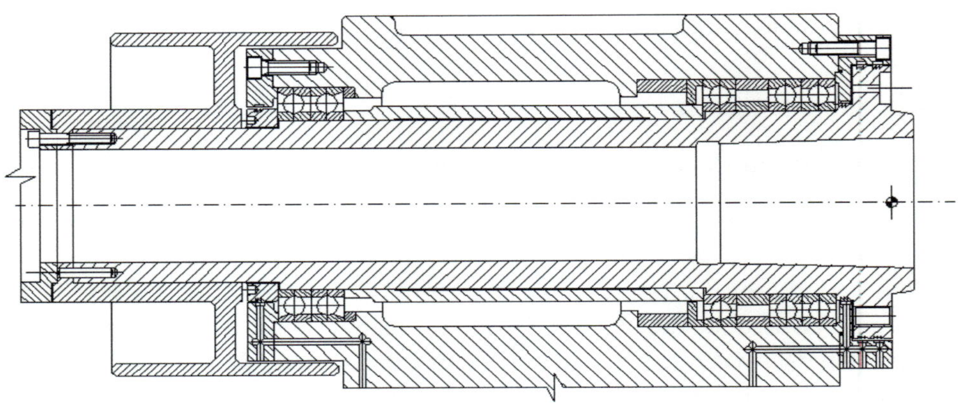

Figure 33. Optimized design of SBL 500

	Unit	Value	Notice [%]
Total axial stiffness C_a	[N/μm]	372	
Total radial stiffness C_r	[N/μm]	351	
Total spindle displacement y_r	[μm]	18,45	
displacement forces resulting from	[μm]		
- the bending moments y_{Mo}	[μm]	9,79	53,0
- the bearing compliance y_L	[μm]	6,16	33,5
- the skidding y_t		2,49	13,5
Limited frequency of rotation n_c	[min^{-1}]	2695	unfit
Life-time T_h	[hour]	5175	unfit
Distance between supports L	[mm]	327	

Table 2. Calculated values of optimized design

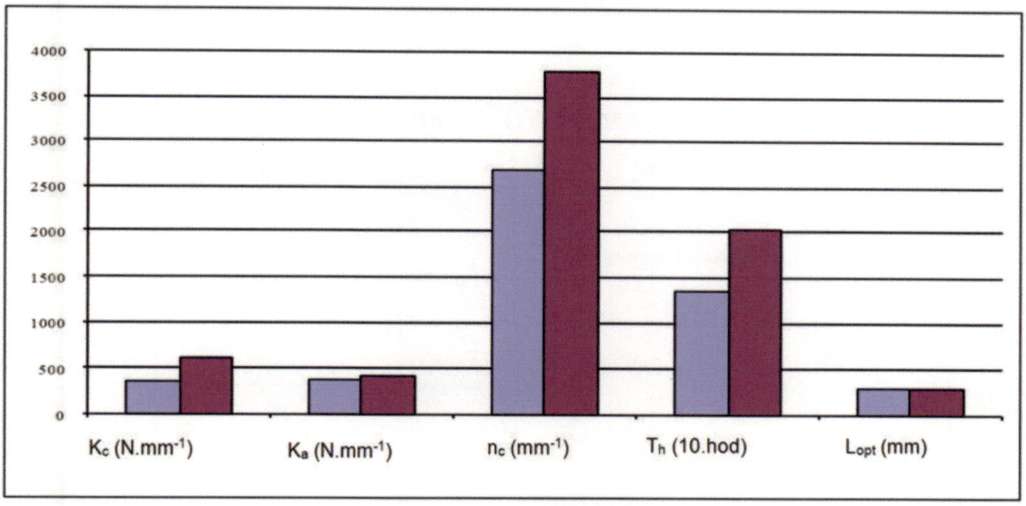

Figure 34. Headstock of SBL 500, Original design, Optimized design

6.2. Dynamic analysis

The most valuable advantage of this system is the possibility of calculating dynamic stiffness at different revolving frequencies of the spindle. The given mathematical model was verified on a number of spindles with programs which enabled the calculation of natural frequencies (COSMOS).The results were in good compliance [24].

The verified spindle, which complied with research findings, was reduced to a three discrete parts. The dynamic mathematical model described above was used to calculate the natural frequencies and dynamic deflections. Table 3 compares calculated and experimental values.

Frequency	Calculated	Experimental	Difference
f_1 (Hz)	1 201	940	+27,8 %
f_2 (Hz)	1 727	1610	+7,3 %
f_3 (Hz)	10 605	-	-

Table 3. Experimental and calculated values of frequencies

The results can be considered as correct, in spite of the relatively large difference in values (28 %) in the first frequency. This is as a result of the fact that the dimensions of the additional rotating parts are not included. If these parts were included in the calculation, the values of the calculated natural frequencies would be smaller.

An example of the graphic output of calculated values is shown in Figure 35, [23]. The chart shows the dynamic deflection of the spindle reduced to three masses. The first two resonant frequencies of the optimized spindle are marked on the chart.

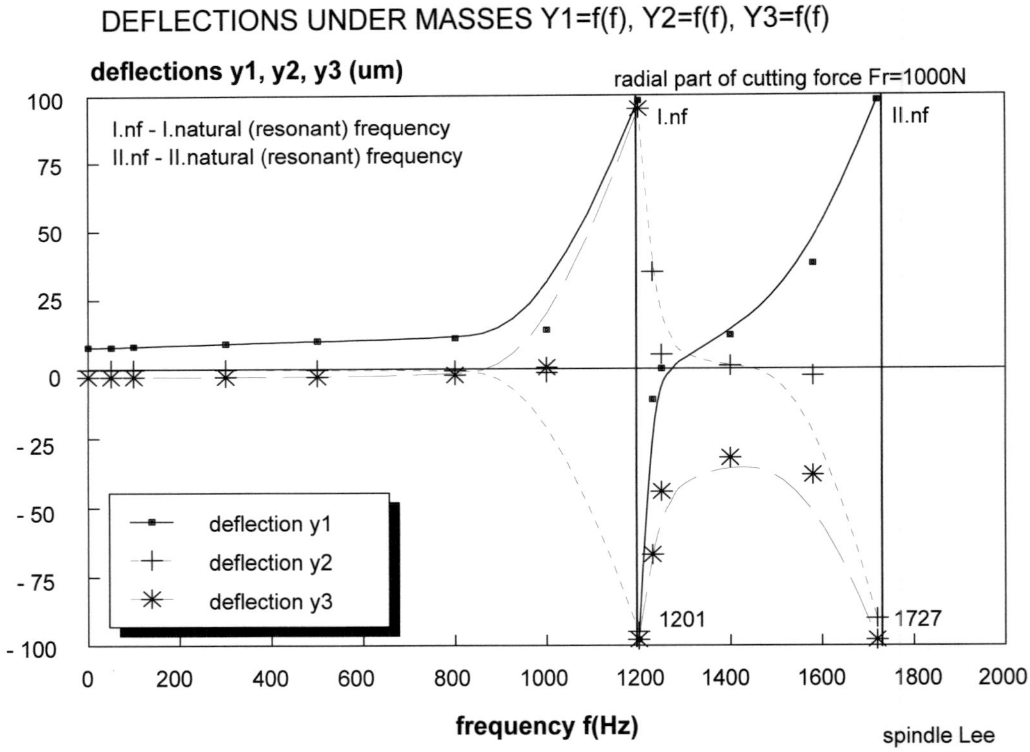

Figure 35. Dynamic deflections of the spindle according to research findings [7]

6.3. Conclusion

One of the main requirements in designing new spindle housing systems is the ability of the design to be quickly applied to real world practice. The methodologies of calculation that were created must be verified, and models must be adapted into a suitable user friendly, computerized format. The models must illustrate the real characteristics of a spindle housing system.

In this design process, only one variable or parameter was changed and the optimal configuration was identified. The results calculated for a static analysis of the SBL Headstock are presented in Table 2 and Figure 33. The dynamic analysis results are presented in Table 3 and Figure 35. The calculated results were verified with experimental measurements. The difference between measured and calculated values is relatively small.

There is no doubt that the re-design has been a success story, and has proven to be highly effective in the identification of optimal SBS design. More detailed information can be read in [22], [23] and its application can be seen in the machine tools made by TRENS Inc., The SBL Lathe was presented in the Mechanical Engineering Exhibition in Nitra in 2010 and in the EMO Exhibition in Düseldorf in 2011.

7. Nomenclature

N - high-speed ability
δ – elastic compression
F – external load
P - roller loading
E - modulus of elasticity of the material
J – quadratic moments of inertia
i – number of bearings
α – contact angle
D, d – diameter
n – high spindle revolutions
l – distance of curvature centre
K - stiffness
γ – pitch angle
O – centre

INDEXES

a – axial direction
r radial direction
z – built-in state
1 – referring to its bearing
0 – roller loaded to the maximum

j – circumference roller
A – external ring
I – internal ring
m – medium value
w - roller bearing

Author details

Ľubomír Šooš
STU Bratislava, Institute of manufacturing systems, environmental technology and quality management, Bratislava, Slovakia

8. References

[1] Weck, M., - Hennes, N. - Krell, M. (1999): Spindlel and Toolsystems with High Damping. In: Cirp Annals-manufacturing Technology - CIRP ANN-MANUF. TECHNOL , vol. 48, no. 1, pp. 297-302, 1999..

[2] Marek, J a kol.: Konstrukce CNC obráběcích strojú. MM Publisching, s.r.o., Praha 2010, ISBN 978-80_254-7980-3, 419 s p.

[3] Lee, D. - Sin, H. - Sun, N. (1985) Manufacturing of a Graphite Epoxy Composite Spindle for a Machine Tool. CIRP, 34, number 1, pp. 365 -369.

[4] Šooš, Ľ. (2008) Spindle headstock - the heart of machine tool. In: Machine Design: On the occasion of 48th anniversary of the Faculty of Technical Sciences: 1960-2008, Novi Sad: University of Novi Sad. ISBN 978-86-7892-105-6., pp. 335-340.

[5] Šooš, Ľ. (2008) Quality of design engineering: Case of machine tools headstock. In: Quality Festival 2008 : 2nd International quality conference. - Kragujevac, May 13-15, 2008. Kragujevac: University in Kragujevac, ISBN 978-86-86663-25-2.

[6] Šooš, Ľ. (2008) Contribution to the research of static and dynamic properties of CNC turning machine In: Strojnícky časopis = Journal of Mechanical engineering. ISSN 0039-2472. - Roč. 59, č. 5-6 (2008), pp. 231-239

[7] Javorčík, L. - Šooš, Ľ. - Zon, J. (1991) Applied software technology for designing a bearing housing fitted with rolling bearing arrangemant. in.:"ICED 91". Zurich, August, 1991. pp.1228-1233.

[8] Šooš, Ľ. – Javorčík, L.- Šarkan, P. (1995) An inteligent drive unit for Machine Tools. In.: The first world congress on Intelligent manufacturing porceedings, university of Puerto Rico, 13-17. February 1995, pp. 344-352.

[9] Demeč, P. (2001) Presnosť obrábacích strojov a jej matematické modelovanie. - 1. vyd. - Košice: Technická univerzita v Košiciach. - 146 p. - ISBN 80-7099-620-X, (in Slovak).

[10] Šooš, Ľ. (2008) Criteria for selection of bearings arrangements. In: 32. Savetovanje proizvodnog mašinstva Srbije sa medunarodnim učešcem = 32nd Conference on production engineering of Serbia with foreign participants: Zbornik radova =

Proceedings. - Novi Sad, 18.-20. 9. 2008. - Serbia: Fakultet tehničkih nauka. - ISBN 978-86-7892-131-5. - 395-399 p.

[11] Šooš, Ľ. (2010) New methodology calculations of radial stiffness nodal points spindle machine tool. In: International symposium on Advanced Engineering & Applied Management - 40th Anniversary in Higher Education: Romania /Hunedoara/ 4-5 November, 2010. - Hunedoara: Faculty of Engineering Hunedoara. - ISBN 978-973-0-09340-7. - III-99 - III-104.

[12] Šooš, Ľ. (2011) Approximate methodology calculations of stiffness nodal points. In: World Academy of Science, Engineering and Technology. - ISSN 2010-376X. - Year 7, Issue 80, pp. 1390-1395.

[13] Harris, T.A.: (1966) Rolling Bearing Analysis. New York - London - Sydney, 1966, 481 ps.

[14] Balmont, V.B. - Russkich,S.P.: (1978) Rasčet radialnoj žestkosti radialno - upornogo podšipnika. Trudy instituta. M., Specinformcentr VNIPPa, 69, 1978, č.1, s..pp. 93 - 107.

[15] Kovalev, M.P.- Narodeckiij,M.Z.:(1980) Rasčet vysokotočnych šarikopodšipnikov. 2 vyd. Moskva, Mašinostroenie 1980. 279s p.

[16] Šooš, Ľ. (2008) Radial stiffness of nodal points of a spindle. In: MATAR Praha 2008. Part 2: Testing, technology: Proceedings of international congresss. - Prague 16th-17th September, Brno 18th September 2008. - Praha: České vysoké učení technické v Praze - ISBN 978-80-904077-0-1. - pp. 43-47.

[17] ŠoošŠOOŠ, Ľ.- BÁBICSábics,J.: (1989) Axiálna tuhosť vysokootáčkových vretien obrábacích strojov. In.: Strojírenství, 39, 1989, č.2, pp s. 86-91.

[18] Šooš, Ľ. - Šarkan, P. (2004) Design of spindle –bearing arrangement of angular ball bearings. In.: MMA 94: Fleksibilne technologije: 11th International conference on Flexible Technologies. Novi Sad, 8 – 9.6.2004. - Novi Sad: Institut za proizvodno mašinstvo -pp. 271-275.

[19] Šooš, Ľ.- Valčuha, Š. - Bábics, J.: PV 08651-88 Zariadenie na skúšanie valivých ložísk [Patent].

[20] Šooš, Ľ.: Generátor skladaných rotačných pohybov. - 2009. - Číslo úžitkového vzoru: SK 5363. - Dátum nadobudnutia: 22. 12. 2009, [Patent].

[21] ŠoošŠOOŠ, Ľ.: Duplo pohon, realita alebo vízia? In: Acta Mechanica Slovaca. - ISSN 1335-2393. - Roč. 10, č. 2-A / konf.(heslo) Celoštátna konferencia s medzinárodnou účasťou. 8. ROBTEP 2006. Jasná - Nízke Tatry, 31.5.-2.6.2006 (2006). - Košice : Technická univerzita v Košiciach, spp. 515-518

[22] Šooš, Ľ. Contribution to the research of static and dynamic properties of CNC turning machine. In: Strojnícky časopis = Journal of Mechanical engineering. - ISSN 0039-2472. - Roč. 59, č. 5-6 (2008), pp. 231-239.

[23] Šooš, Ľ.: Spindle - housing system SBL 500 CNC. In: Eksploatacja i Niezawodnošč = Maintenance and reliability. - ISSN 1507-2711. - Č. 2 (2008), pp. 53-56.

[24] Šooš, Ľ. New methodology calculations of radial stiffness nodal points spindle machine tool. In: International symposium on Advanced Engineering & Applied Management - 40th Anniversary in Higher Education: Romania /Hunedoara/ 4-5 November, 2010. - Hunedoara: Faculty of Engineering Hunedoara, 2010. - ISBN 978-973-0-09340-7. - III-99 - III-104.

Chapter 4

Rolling Contact Fatigue in Ultra High Vacuum

Mike Danyluk and Anoop Dhingra

Additional information is available at the end of the chapter

http://dx.doi.org/10.5772/51194

1. Introduction

Rolling elements, such as ball bearings and races, contain surface imperfections known as asperities. The height of surface asperities may be quantified through surface roughness analyses, which assigns a Ra number related the characteristics of the surface and asperities. The depth and width of the valleys between asperities is one significant characteristic of a surface that influences wear, friction, and contact fatigue life of the rolling element.

Surface lubrication may be divided into three categories: i) full film, ii) boundary layer, and iii) mixed film and boundary lubrication (Bhushan, 1999). With full film lubrication, the film is sufficiently thick so that surface asperities do not protrude through the film and will not contact the mating surface. Boundary lubrication describes the condition in which a film is present, but load is transferred between asperity peaks on the surfaces and not the film. Mixed lubrication conditions assume that both film and asperity transmit contact load, and therefore both must be considered in the analysis. A numerical approach capable of modeling all three types using fluid lubrication may be found in (Hu and Zhu, 2000). The proposed model is found to work well over a specific range of film-thickness-ratios and surface RMS roughness.

The need to quickly determine the fatigue life of rolling elements has given rise to rolling contact fatigue test methods that enable fatigue testing at reduced cost. Historically, RCF tests have used petroleum-based liquid-lubricants, which restrict rotational speed of the test due to liquid-lubrication churning. In comparison, RCF testing using solid film lubricants in ultra-high vacuum enables higher rotational speeds leading to test results in less time. For example, a rod composed of a candidate bearing material may accumulate over 10 million stress cycles in a few days running at 130Hz in ultra-high vacuum. The exact number of stress cycles accumulated on the rotating elements will depend on the specifics of the test configuration, such as ball diameter, rod diameter, and the number of balls present. In contrast, RCF testing in air using oil based liquid-lubrication is speed limited, usually to 60 Hz or less, and is limited to a maximum of three ball-contact elements. RCF testing in air using liquid-lubrication requires more time to accumulate the same number of stress cycles.

Rolling contact bearings may have multiple and diverse layers of gaseous molecules accumulated on their surface that can significantly influence friction and wear behavior. Physisorption and chemisorption of external molecules on to the substrate surface effect wear and friction, particularly for thin-films used in high vacuum conditions. Surface texture has a significant influence on wear and friction behavior as well since it determines the nature of contact. Physisorption layers of atoms involve weak Van dar Waals bonds. These layers can begin to detach from the substrate at pressures below 10^{-4} Pa in a process described as surface out-gassing. The energy to initiate physisorption outgassing is roughly 1-2 kCal/mol. Physisorbed layers of atoms do not share electrons with surface atoms and therefore may be easily removed when heat is applied in vacuum conditions. Chemisorbed layers however do share electrons with surface atoms and these bonds can be very strong with an associated energy of 10-100 kCal/mol. Physisoption occurs on all surfaces when exposed to air while chemisorption requires a chemical reaction with the substrate surface and is therefore heavily influenced by surface chemistry.

Hertzian contact analysis considers the deformation of two elastic solid surfaces with the following assumptions: i) the surfaces are smooth, continuous, and nonconforming to each other, ii) the strains are small in the contact area, iii) each solid can be modeled as an elastic half space in the proximity of the contact, and iv) the surfaces are frictionless. The two contacting surfaces can be of general shape, but most often they are chosen to be convex. Maximum shear stress occurs below the surface in the location of the contact. The depth of maximum shear stress is related to the radius of curvature and Poisson ratio of the substrate material. For example, Hertzian contact analysis applied to a material with Poisson's ratio of 0.3 is calculated to have maximum shear stress at a depth of 0.48 times the radius of the contact. Maximum compressive stress occurs at the point of contact. Maximum tensile stress takes place just beyond the edge of the contact area on the surface. Considering RCF loading conditions, subsurface cracking that eventually leads to surface spall and flaking begins at the location of maximum shear stress below the surface of the ball. If a thin film is present, an interfacial spall may result at the location of either maximum compressive or tensile stress, while subsurface cracking will occur within the substrate below the contact zone.

Understanding wear and friction requires knowledge of the type of contact between the two solid bodies. Incipient sliding occurs when two contacting bodies are pressed together such that a stick-point exists within the contact area and materials from each body slide relative to each other about that point. There is no imposed relative motion between the bodies during incipient sliding condition, rather sliding occurs due to elastic deformation of each surface around the stick-point. Friction between rolling contact elements is proportional to the shear strength of the base materials. Stresses in the contact area greater than the yield stress of the material are possible due to compression at the point of contact.

One of the benefits of solid lubricant coatings such as silver, lead, gold, and MoS_2 is their low shear strength to reduce subsurface stresses during incipient sliding. The compliant nature of these solid film lubricants helps to reduce subsurface stress by reducing shear loading in the substrate. However, if increased wear resistance is desired, then a hard and stiff layer is preferred which presents a trade-off when selecting a coating system: soft compliant layer to reduce subsurface stress, or, a hard and stiff layer to increase wear resistance. The effective hardness of a thin solid film system is influenced by the hardness of its layers and substrate along with the elastic-plastic behavior of the coating system. If the film is very thin then the effective hardness is mostly influenced by substrate hardness, leading to an optimized lubricant coating that is thin and soft, applied over a hard-stiff coating bonded to the substrate. The soft lubricant reduces subsurface stress and the hard and stiff coating increases wear resistance.

Solid to solid contact results in two broad categories of surface bonding: adhesive and cohesive. Cohesive bonding takes place in the bulk material and has chemical interactions with covalent bonding, metallic bonding, and ionic and electrostatic bonding. Adhesive bonding involves two dissimilar surfaces and includes physical interactions and Van der Waals bonding. Adhesion is a function of material combination, crystal structure, surface condition, ambient temperature, and crystallographic orientation. Indeed, good adhesion is desired between the solid film lubricant and the surface of the rolling element, and very little cohesion is desired within the solid lubricant itself to reduce friction.

The technology of solid lubricants grew rapidly during the 1960s and 70s to meet the needs of the aerospace industry and for operation in ultra-high vacuum environment. Silver, gold, and lead along with compounds like MoS_2 and graphite were needed for environments in which liquid oil-based and hydrocarbon out gassing could not be tolerated, such as, rotating devices within very high electrical potential or inside space satellite guidance equipment. Test equipment used to gather tribology data for solid lubrication systems were developed through NASA, and industry leaders such as Timken, Du Pont, and General Dynamics to name a few. Platforms were developed for high vacuum and high temperature applications that involved sliding-contact, rolling-contact ball bearings, metal disk and riders, and sliding contact of a sleeve on a cylindrical rod and may be found in (NASA SP-5059, 1972). Each of these test platforms was designed to test a specific load situation.

Solid lubricants find unique application in systems with an operating pressure range of 10^{-4} to 10^{-9} Pa and are also temperature resistant in high vacuum. Rotating anode x-ray tubes that use ball bearing elements are one example of a rolling contact configuration under high vacuum and at high temperature. Non-volatile thin-film lubrication is required due to operation in vacuum and in the presence of high electrical potential. Solid lubrication systems are greatly influenced by process history and operating conditions. It is strongly suggested that testing and evaluation involving any solid film be done using equipment known to correlate with the specific application for which the coating system is being considered. RCF testing of coating systems and ball-bearing materials is one example of a test method that closely simulates rolling element bearing systems.

1.1. Current research

Matthews et al. (2007) report that thin film coating survival is most dependent on adhesion and substrate subsurface bonding. Surface studies at the nano scale suggest that friction begins at the atomic level through mechanical vibrations within the lattice structure of the substrate. Substrate material properties influence spall behavior and RCF life as well. Micro-tribology considers the mechanics of cracks and fracture in the material at the asperity level. Berthier et al. (1989) consider velocity differences between contact surfaces and suggests four mechanisms to accommodate these differences: elastic deformation, fracture, shear stress, and rolling contact. They report that although hard coatings are preferred for wear resistance, the addition of a thin compliant film that includes nano scale hard elements in the solid lubricant reduces friction as well.

Sadeghi et al. (2009) report that subsurface failures usually occur during RCF testing due to fatigue. If a rolling element such as a ball-bearing is kept lubricated and clean, then the primary failure mode will be subsurface spall from fatigue. Fatigue failure beneath the substrate surface is thought to proceed as follows: i) work hardening, ii) elastic response, iii) material softening leading to yield. In a solid lubrication system that uses silver for example, work hardening of the substrate occurs within the run-in period in addition to silver transfer between the surfaces. If there is a sufficient amount of silver, the test will operate in the elastic response regime for as long as the silver is present. Material softening results in a larger volume of material that has yielded plastically. Plastic yielding begins the onset of subsurface cracking which then propagates to the surface resulting in a spall. Cracks and the onset of spall may also originate from inclusions or material defects, which, helps to explain the huge scatter in RCF life test data.

Polonsky et al. (1998) confirm two types of RCF failures related to composite coatings: subsurface-initiated spall and near-surface coating failure. They report coating-cohesive failure of 750 nm thick TiN running against softer 12.7 mm diameter ANSI M50 steel balls due to interface initiated spall in the TiN coating. Rosado et al. (2010) sought more understanding of the spall growth process once a spall has occurred and identify material parameters to optimize and improve spall growth resistance. For contact stress of 2.41 GPa, they report slower spall growth in ANSI M50 compared with either ANSI M50NiL or ANSI 52100 due to material composition and processing. Danyluk and Dhingra (2011) observe two types of failures related to RCF testing in ultra-high vacuum: coating depletion and surface spall. Solid film nickel-copper-silver lubrication was applied to ball bearings over a range of process voltage and pressure using a physical vapor deposition ion plating process. Coating depletion and surface spall failures were the two most common failure modes, and a unique spall-related non-precession type failure was also observed. The non-precession failure mode was associated with higher process voltages and with reduced RCF life.

Higgs and Wornyoh (2008) use conservation of mass to model an in-situ mechanism for self replenishing powder lubrication on sliding contacts. Application of Archard's wear law and a third-body concept are used to formulate the conservation mass equation. The control-volume-fraction-coverage (CVFC) model concept is applied to set the bounds of their theoretical approach. Wornyoh and Higgs (2011) extend the work of (Higgs and Wornyoh, 2008) to an asperity-based fractional coverage (AFC) model derivation and analysis. The latter AFC model extends the surface applicability to tribo-surfaces using atomic force microscopy that enable inclusion of surface topology information in the film transfer model.

1.2. Chapter overview

A review of the current literature indicates that ball-on-rod RCF testing in high vacuum and at high speeds (approaching 130 Hz) has not received much attention; the work presented here is meant to fill that gap related to high speed RCF testing of tribology coatings. A third-body approximation that accounts for coating wear has been implemented as well for aid to predict test life based on contact stress. A Lundberg-Plamgren empirical model is also presented for comparison with the third-body approximation model. Surface and coating diagnostic tools are presented to show the relationship between coating composition and RCF life.

2. Experimental procedure

The RCF test rig in Figure 1 was assembled using off the shelf components purchased from leading vacuum and mechanical equipment vendors (Danyluk and Dhingra, 2012a). For example, a Kollmorgen™ servo motor (model number AKM21E) is used to rotate the rod. The drive motor is mounted underneath the chamber and motor torque is delivered to the test rod inside the vacuum chamber using a ferro-fluidic rotary feed-through device similar to Kurt J. Lesker Vacuum™ part number FE121099. High vacuum is applied using a Varian™ V-81M turbo pumping system as shown in the right panel of Figure 1.

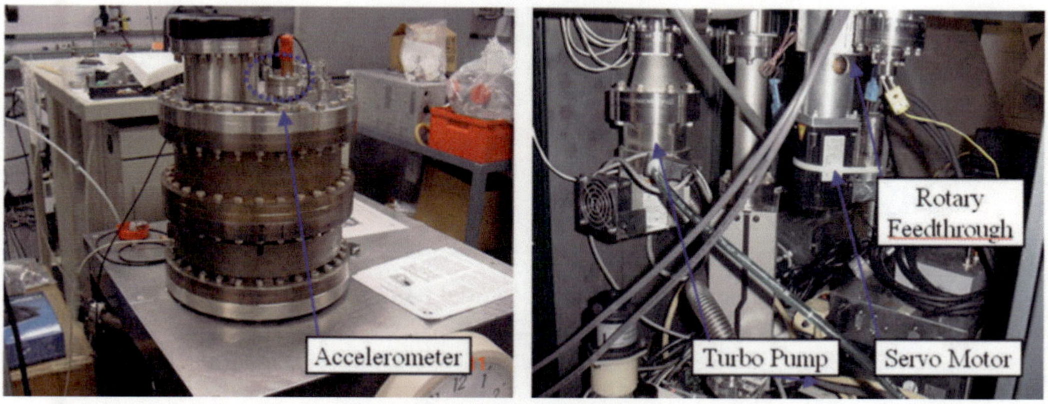

Figure 1. Support hardware and UHV-RCF chamber: turbo pumping system, residual gas analyzer, servo motor drive, and ferro-fluidic rotary feed through.

The RCF hardware inside the chamber is a modification of the three-ball-rod test rig of reference (Hoo 1982). The test-head hardware components were fabricated from 304L stainless steel and are positioned inside the UHV chamber as shown in Figure 2c. A component view of the test chamber is shown in Figure 2a, with emphasis on hardware type and assembly order. The RCF test elements: the balls, rods, and races, are shown in Figure 2b. The races are press fit into the test fixture and held stationary while the balls rotate between the fixed races and the rotating rod. A typical test consumes 5 or 6 balls depending on which ball size is used: 12.7 mm or 7.94 mm. For example, when testing with 12.7 mm diameter balls as shown in Figure 2c, five balls, two races and one rod will constitute a test.

If 7.94 mm balls are used then six balls are required. A description of test element combinations that may be carried out using this platform is presented in Table 1.

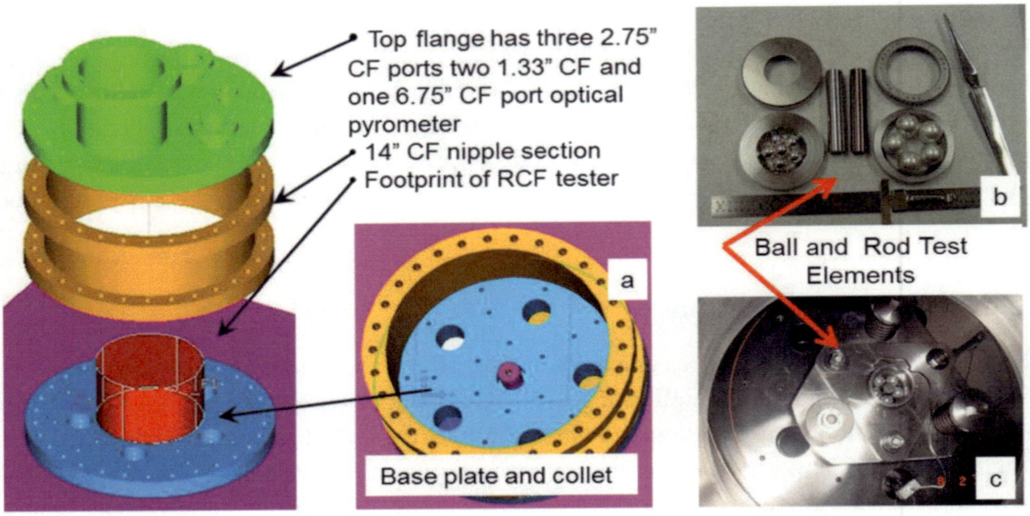

Figure 2. RCF test chamber and hardware: a) chamber sections assembly order, b) Rex 20 and M50 balls, races and rods, c) M50 balls against a Si_3N_4 rod with M50 outer races inside the vacuum chamber.

Referring to Table 1, the Si_3N_4 rods were purchased from Ceradyne™ Inc, and were processed using the EKasin™ method from Ceradyne. These rods were chosen for their high temperature and high thermal shock resistance. The Rex20 rods and races, along with the 12.7 mm diameter M50 steel balls were purchased from Timken™ and are used extensively for RCF testing with oil based lubrication. The 7.94 mm diameter ANSI T5 balls were purchased from Koyo™.

Config.	Ball Material	Rod Material	Race Material	Lubrication	Ball Size (mm)	Number of balls
1	M50 Steel	Rex 20	M50 Steel	Silver	12.7	5
2	ANSI T5	Rex 20	Rex 20	Ni/Cu/Ag	7.94	6
3	M50 Steel	Si_3N_4	M50 Steel	Silver	12.7	5

Table 1. Test configurations for RCF testing of test elements shown in Figure 2b in ultra-high vacuum.

2.1. Test material measurements

Coating survival and RCF test life is dependent on coating and substrate material properties. For example, a thin film deposited onto an M50 Steel substrate is likely to spall if the applied contact stress during the RCF test exceeds the ultimate strength of the substrate. When considering the ultimate strength and hardness of rolling bearing elements it is best to directly measure surface hardness and then approximate the ultimate strength. Substrate geometry influences surface hardness and it is to be expected that ball, rod, and race geometries will influence the hardness measurement as well. Table 2 presents hardness and material property data for the test elements presented in Figure 2b. Table 3 contains measured surface roughness data. When collecting hardness and surface roughness data, it is recommended to carryout repeated measurements over multiple samples. For example, the data presented in Tables 2 and 3 were repeated three times per sample on five samples and the average was taken based on 15 measurements of each test element.

Material Property	9.53 mm Diameter Rex 20 Rod	9.53 mm Diameter Si_3N_4 Rod	7.94 mm Steel ANSI T5 ball	12.7 mm Steel M50 ball	Rex 20 Race	M50 Race
HRC (measured)	63.2	74.7	61.3	61.9	65.8	44.1
Elastic Modulus (GPa)	235	310	214	203	235	203
Poisson Ratio	0.29	0.25	0.29	0.29	0.29	0.29

Table 2. Hardness and material property data for RCF test elements in Figure 2b.

12.7 mm M50 Ball	M50 Race	7.94 mm T5 Ball	Rex20 Race	9.53 mm Si_3N_4 Rod	9.53 mm Rex20 Rod
0.31	0.32	0.04	0.15	0.05	0.11

Table 3. Average surface roughness Ra data in microns for RCF test elements.

2.2. Test preparation

Preparation and process history of the test elements have significant influence on RCF life. Thin film coatings on the order of 500 nm thick are very susceptible to surface contamination from exposure to air, or contact with volatile substances such as organic compounds along with low melting temperature metals such as tin or indium for example. Component cleanliness and honest adherence to good vacuum procedures are required to ensure validity of test results. For example, prior to silver deposition all of the balls were cleaned in an ultrasonic bath in methylene chloride for 20 minutes to remove oil and particulates. Prior to coating, the balls were also outgassed inside a UHV chamber at 10^{-7} Pa for 24 hours and then scrubbed with argon plasma before approximately 200 nm of silver was deposited on to the ball surface. Steps were taken to insure even coating thickness during deposition. The rods and races were cleaned similarly and all test components were stored in warm dry nitrogen. Pre-coated T5, 7.94 mm diameter balls that were purchased from Koyo were vacuum-packed and then stored in warm dry nitrogen until loaded into the test rig. Each time the test elements are exposed to air an out-gassing procedure should be applied prior to the start of the test. For example, after the balls and rod were loaded into the UHV-RCF test chamber, the system was outgassed for 12 hours at 10^{-7} Pa prior to starting the test.

2.3. Failure criterion

Based on previous testing using the RCF rig in Figure 1, the onset of a spall on at least one ball is detectable over a vibration range of 0.22 g to 0.35 g with the accelerometer mounted on the top section of the test chamber. To be clear, each test chamber may have different vibration transmissibility based on its assembly. Therefore a new test rig should be characterized for the first detectable vibration of the onset of coating spall, and verified by post-test autopsy. The failure threshold for the rig presented in Figure 1 is 0.35 g, as measured by an accelerometer mounted on the top section. For the first 15 minutes of the test, a run-in period occurs and the vibration can reach 0.30 g, then settles to a steady state range of 0.06 g to 0.15 g for length of the test until failure. The test fixture temperature measurement shown in Figure 2c tends to increases with vibration and in general tended to increase over the length of the test. This is to be expected since as the test proceeds the solid lubrication is depleted and there is an increase in surface friction and ultimately increased vibration. Temperature at failure may vary based on coating system and contact stress loading. For the test rig shown in Figure 2c a thermocouple has been placed in contact with the top race to monitor temperature during the test. An optical temperature measurement may be taken as well from the large port in the top section.

2.4. Thin film diagnostics

Confirmation of thin film composition and thickness prior to RCF testing can give insight and predictability to the test results. RCF test results may be confounded due to unknown and unwanted constituents within the coating. Auger Electron Spectroscopy (AES) may be used to measure and sample atoms within the coating. The AES process uses a high energy electron beam to bore a small diameter hole, on the order of 1 – 2 nm, into the coating and ball surface. The material that is removed during the process is analyzed using an in-situ mass spectrometer to determine its species. Figure 3 presents constituent information related to a thin film of silver that was deposited onto a 7.94 mm diameter ANSI T5 steel ball. Starting from the left, which correlates to the surface of the coated-ball, carbon, oxygen, and silver are present in the coating. Moving to the far right in Figure 3, there is a strong transition from silver to iron and chromium, which are two constituents of T5. Based on the known composition of T5 and the AES results of Figure 3 the coating thickness may be approximated as 190 nm. More interesting however is the composition of the coating through its thickness, specifically, the iron, oxygen, and nickel present throughout the silver layer. The results of Figure 4 illustrate element composition of a nickel-copper-silver coating deposited on to a Si_3N_4 ball using a physical vapor deposition ion plating process. Referring to Figure 4, there is a high concentration of nickel and iron near the ball-coating interface at about 120 nm. Since the ball itself does not contain Ni or Fe, the presence of those elements and their concentrations suggest that contamination occurred during the deposition process. The results of Figures 3 and 4 suggest interlayer mixing and coating contamination during the deposition process which is likely to influence the lubrication properties of the film, and ultimately the RCF life. For more information concerning thin film diagnostics and deposition plasma diagnostics related to RCF life, see (Danyluk and Dhingra, 2012b).

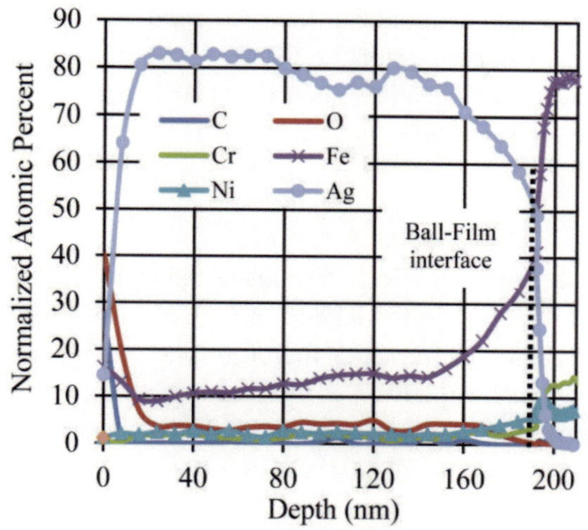

Figure 3. Auger electron spectroscopy depth profile of a silver film deposited on a 7.94 mm diameter ANSI T5 steel ball.

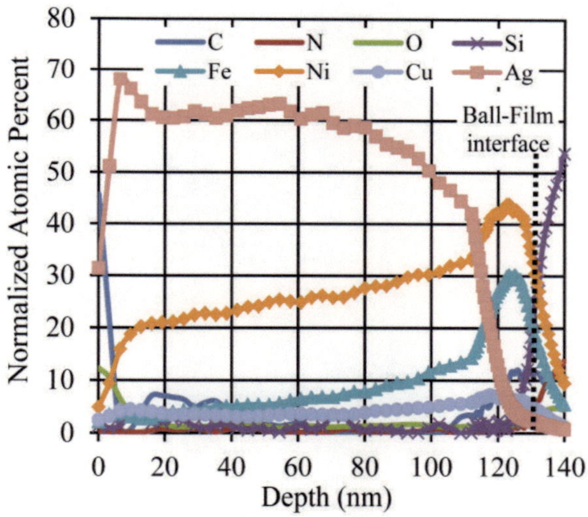

Figure 4. Auger electron spectroscopy depth profile of a nickel-copper-silver film deposited on a 7.94 mm diameter Si_3N_4 ball.

3. Experimental results

Statistical tools may be used to extract necessary information from RCF data. For example, Weibull analysis and high-cycle fatigue software tools such as Reliasoft™ may be used to correlate and compare test results independent of the coating and test elements in the test. An inverse power law life model with Weibull distribution may be used to check the experimental setup by comparing the Weibull parameters against known test configurations. Historically, a Weibull shape factor in the range $1 < \beta < 4$ is to be expected for bearing and gear type RCF failures. For comparison, a shape factor less than 1 would indicate a flawed test method or infant failure. It is good practice to fit the RCF data to a Weibull distribution model starting after the fifth test so that one may confirm right away that the test results reflect coating and material performance and not a flawed assembly process or inadequate test preparation.

3.1. RCF statistical data

Cycles versus contact-stress for two RCF test loads using configuration 3 is presented in Figure 5. The shape factor shown in the legend of Figure 5 is within the expected range for bearing and gear type RCF testing and therefore the results reflect true film/coating performance. An eta-line has been added as well for extrapolation to testing at stress levels between the data presented in Figure 5. The shaded areas represent the fitted probability density function (PDF) based on the RCF data at that stress level.

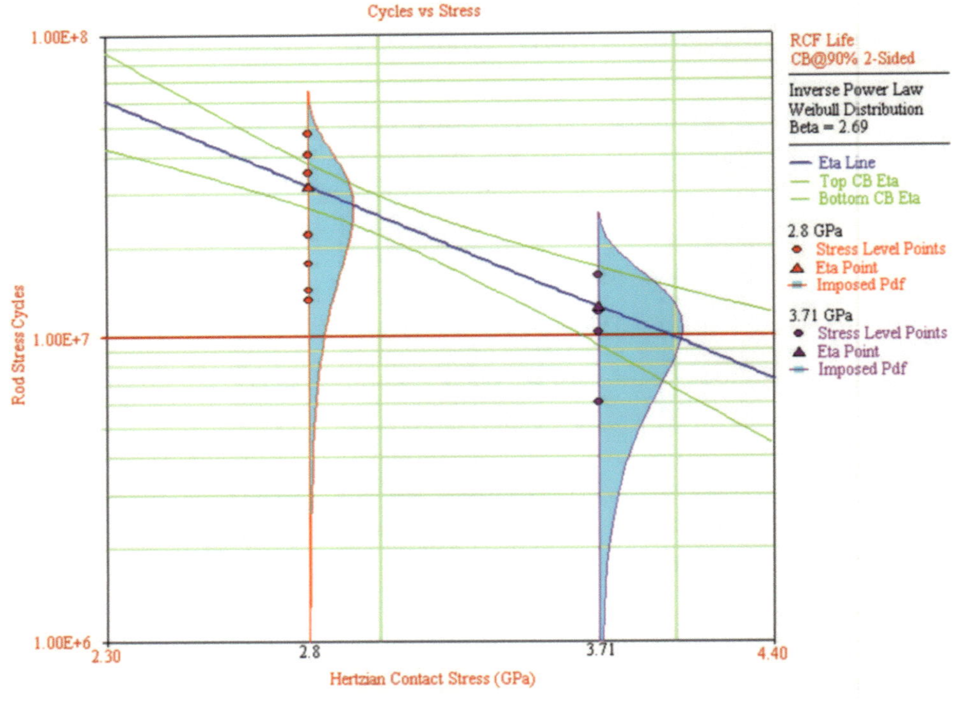

Figure 5. Cycles versus stress for 12.7 mm M50 steel balls with M50 races against a Si_3N_4 rod. Test rotation of 130 Hz in high vacuum with approximately 200 nm of silver on the balls.

Figures 6 and 7 contain reliability prediction information for two thin film systems: evaporated silver and, ion plated nickel copper silver. The Weibull shape factor for each set of data in Figures 6 and 7 is shown in the upper right corner as 2.54 and 2.76, respectively. These beta-factors are within the expected range for bearing RCF testing and therefore the data is an accurate representation of coating and rod performance. The stress-use parameter used in the reliability calculation is also presented in the figures. The stress-use parameter is a modeling tool used to extrapolate reliability data to a loading other than those actually tested. When choosing the stress-use parameter, make sure the data is collected over a sufficient range of contact stresses that includes the stress-use value. The data presented in Figures 6 and 7 was taken over a stress-use range of 1.4 to 3.5 GPa. The results in Figures 5 through 7 strongly suggest rolling contact fatigue failure since the contact stresses for each of these tests are 1/3 less than the calculated tensile yield strength of each component as calculated from the hardness measurements presented in Table 2.

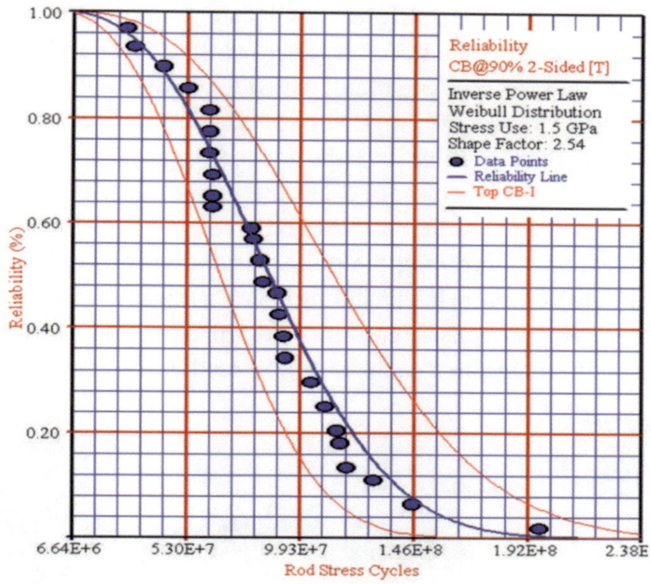

Figure 6. Reliability data for RCF elements in configuration 1 of Table 1.

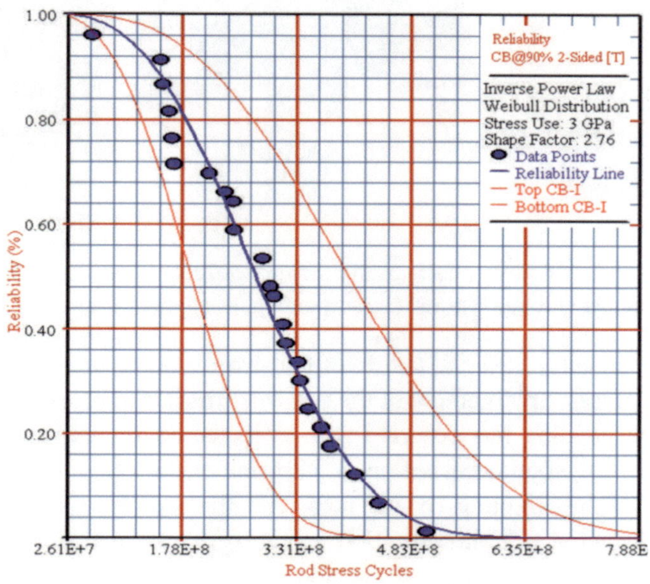

Figure 7. Reliability data for RCF elements in configuration 2 of Table 1.

3.2. RCF post test autopsy

Figures 8 through 10 present post-test photos of the ball, rod, and races for two types of failures and for two RCF loads. Configuration 1 of Table 1 was used for all three tests. Figure 8 contains results from a suspended test after 1.7 hours. A ball-rod-race model is presented in the center panel of Figure 8 to highlight the silver and the wear tracks on the ball and rotating rod. Inspection of the ball surface reveals that silver is still present on the ball surface, as illustrated by the scratch-test results shown in the right panel of Figure 8. Inspection of the rod surface shows traces of silver transfer to the rod wear track. Figure 9 contains results from a life test that eventually failed for silver depletion after 20.6 hours of testing, corresponding to 9.6 x 10^6 stress cycles accumulated on the rod wear track. Information related to silver transfer and the movement of silver on the wear track is highlighted as in Figure 9. The significance of silver transfer from the ball surface to both the rod and race wear-tracks suggests that a third-body-transfer model may be used to model the UHV-RCF test method. Concerning the rod and race information in Figure 9, the solid silver lubricant was depleted by incipient sliding between the rod-ball and race-ball contacts such that all of the silver was transferred out of the third-body storage areas. For the RCF platform, the third-body storage areas may be represented on the rod and race wear tracks.

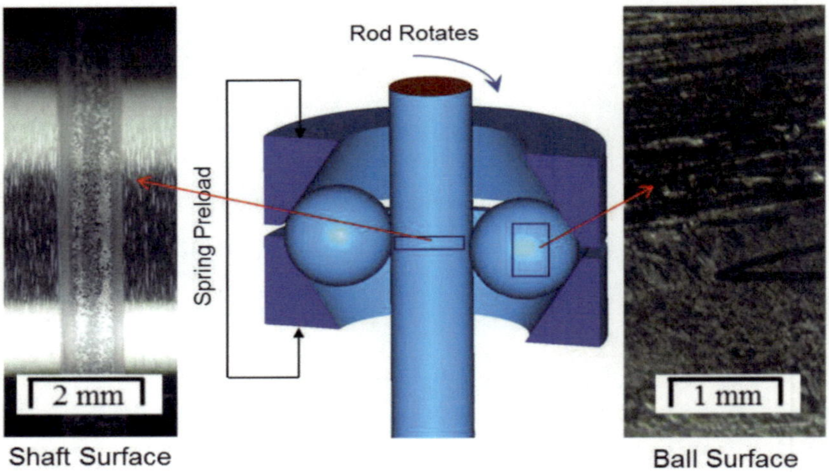

Figure 8. Autopsy data for RCF elements in configuration 1 of Table 1 operated at 3.61 GPa loading. Suspension after 1.7 hours of rotation at 130 Hz, accounting for approximately 7.8 x 10^5 rod stress cycles.

Figure 10 contains an early life spall failure of the silver coating and subsequent damage to the ball and rod surfaces. Enlargement of the rod wear track is to be expected for this failure mode since plastic yielding has occurred due to increased friction associated with spall failures. Once a coating spall occurred on at least one of the balls, the surface is damaged due to yielding and the resulting contact area with the rod increases rapidly. The test is halted after the stopping criteria, vibration levels in excess of 0.35 g, is exceeded for 1 minute. With the rod rotating at 130 Hz and a deceleration time of about 3 seconds, approximately 390 rod-rotations will be added to the failed surfaces before the rod stops rotating. The debris from the spall remains within the wear track resulting in significant plastic deformation of all contact surfaces before the test is stopped. Coating spall failure resulted in higher friction and stress leading ultimately to surface yielding. The silver depletion failure mode resulted in increased friction and vibration as well but without surface yielding before the test was stopped. However, if allowed to continue without lubrication the increased friction would accelerate the onset of subsurface spall of the ball and rod.

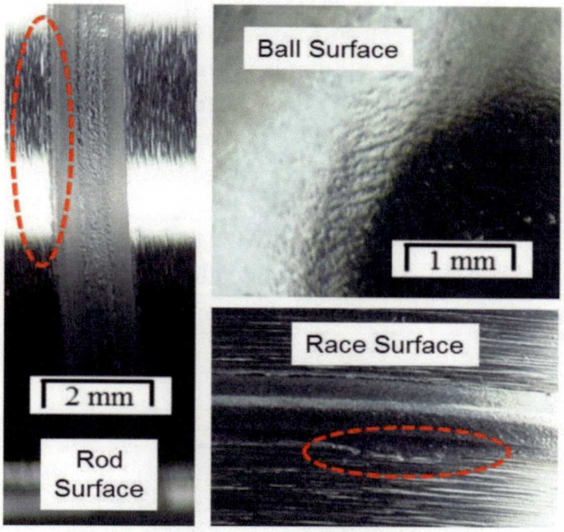

Figure 9. Autopsy data for RCF elements in configuration 1 of Table 1 operated at 3.61 GPa loading. Silver depletion failure after 20.6 hours of rotation at 130 Hz, accounting for approximately 9.6 x 10^6 rod stress cycles.

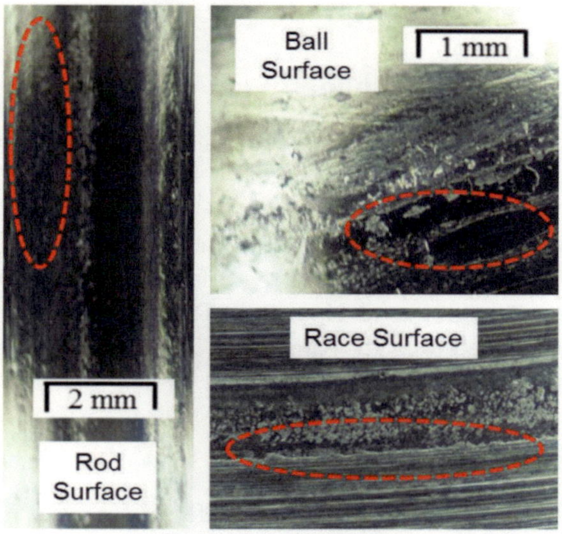

Figure 10. Autopsy data for RCF elements in configuration 1 of Table 1 operated at 4.0 GPa loading. Coating spall failure after 9.1 hours of rotation at 130 Hz, accounting for approximately 4.2 x 10^6 rod stress cycles

Post-test autopsy and Scanning Electron Microscopy (SEM) results may be used to approximate the amount of silver remaining on the ball surface at the end of each test. Table 4 presents element composition of one non-coated and two silver coated balls after testing. These balls account for three types of SEM test results: before coating, after silver depletion failure, and after early spall failure. The surface composition of each ball was derived from energy spectroscopy analyses attached to the SEM instrument. Concerning Table 4, post-test SEM data from a silver depletion failure is shown in column 3 with no silver present. All of the silver was transferred from the ball surface on to the rod and race surfaces, representing termination of the third body transfer mechanism suggested in (Higgs and Wornyoh, 2008). The results of column 4 however show some silver was still present on the ball surface following early spall failure after 9.1 hours. A theoretical analysis for a third-body transfer model using UHV-RCF data collected from suspended tests is presented in the next section.

	Baseline M50 Bearing Steel	SEM 12.7 mm M50 ball, silver depletion failure.	SEM 12.7 mm M50 ball, early spall failure.
Carbon	0.97		
Chromium	4.18	1.58	2.07
Iron	88.60	97.84	77.83
Manganese	0.32		
Molybdenum	4.45		
Silicon	0.27	0.48	0.68
Vanadium	1.21		
Silver	0.00		16.37

Table 4. Element composition of one non-coated ball and two coated balls after testing. Composition derived from energy dispersive spectroscopy using a SEM.

4. Thin film solid lubrication modelling

Test life comparison using two modeling approaches is the focus of this section. The first model uses a conservation of mass approach and is based on the work presented in (Higg and Wornyoh, 2008), (Danyluk and Dhingra, 2012a). A third-body mass transfer concept is applied to account for the transport of the film from the ball surface to the rod and race contact surfaces as seen experimentally in Figures 8 and 9. The second modeling approach is similar to the Lundberg-Palmgren model in that the RCF data from the test configurations of Table 1 are used to fit a load-capacity parameter, C, to L_{10} data similar to that found in chapter 8 of (Bhushan, 1999).

4.1. Third body transfer model

The third-body model concept involves mass transfer of the solid film lubricant from its source to the contacting surfaces within the rolling-contact system. For example, if a specimen of silver that is stationary is pressed into contact with a rotating shaft the resulting contact will enable transfer of the soft silver from the specimen to the contact surface of the shaft. The rate of silver transfer is related to: i) the force of the contact, ii) the surface roughness and speed of the shaft, and iii) the rate at which excess silver is pushed out of the contact area. Concerning the RCF tests in Figures 8 through 10, the source of the solid lubricant is the amount of thin-film coating on the balls at the beginning of the test, approximately 200 nm. The film lubricant is transferred to the rod and race contact areas, which are considered the third-body volumes. Depending on surface roughness, more or less lubricant may accumulate in the valleys between surface asperities on the rod and race surfaces. The solid lubricant on the ball surface represents the source or input to the third-body concept.

The control volume fraction coverage model (CVFC) has been presented and explained in (Higgs and Wornyoh ,2008), however, some parts of that formulation are presented here for clarity. The assumptions of the CVFC model for solid lubrication transfer to the third-body volumes are as follows: i) the ball/rod and ball/race contact surfaces are flat within their contact areas, ii) incipient sliding occurs between surfaces due to elastic deformation, iii) the fractional response and friction of the interfaces is primarily a function of the amount of silver present in the third-body volumes of the race and rod, and on the surface of the ball.

A conservation of mass formulation for the transfer of film lubricant from the ball surface to the wear tracks of the race and rod is as,

$$\begin{pmatrix} \textit{Third Body} \\ \textit{Storage Rate} \end{pmatrix} = \begin{pmatrix} \textit{Third Body} \\ \textit{Input Rate} \end{pmatrix} - \begin{pmatrix} \textit{Third Body} \\ \textit{Output Rate} \end{pmatrix}. \tag{1}$$

The output rate in equation (1) is driven by the load between the ball-rod and ball-race that forces some of the solid silver out of the wear track. Examination of the wear tracks on the races and on the rod and race in Figures 8 through 10 illustrate that silver is pushed outside of the CVFC volume over time, and hence removed from the third-body storage volumes. The input rate to the third-body storage volumes of the rod and race contact zones is influenced by the fiction coefficient between the solid lubricant and contact area. Concerning RCF contact and Figures 8 through 10, incipient sliding between the ball-rod and ball-race is assumed throughout this formulation.

Equation (1) may be described as the rate of change of the fractional coverage, $X(t)$ of the third-bodies on both the rod and race wear tracks. For the present study, $X(t)$ will be normalized to the average surface roughness of the race and rod as presented in Table 3, or approximately 250 nm and represents the maximum asperity height defined as, h_{max}. The asperity depth is about the same as the initial silver coating thickness on the balls as well, approximately 200 nm. Following the form of (Higgs and Wornyoh, 2008), and (Danyluk and Dhingra, 2012a) the fractional coverage variable is defined as,

$$X = \frac{h}{h_{max}}, \tag{2}$$

where h is the local height of silver coating in the third-body volumes. Archard's volume wear rate law is used to account for surface wear interactions and is defined as,

$$\frac{dV}{dt} = KF_N U, \tag{3}$$

where V, K, F_N, and U are the volume, wear coefficient, normal force, and sliding velocity, respectively. The wear coefficient K is the probability that a surface is being worn due to sliding contact, and for this section incipient sliding assumed. Combining equations (1) through (3) gives the following differential equation for $X(t)$ as,

$$Ah_{\max}\frac{dX}{dt} = \left(K_{bc}F_cU_c + K_{br}F_rU_r\right)\left(1 - X\right) - \left(K_{bEc}F_cU_c + K_{bEr}F_rU_r\right)X, \tag{4}$$

where the first term on the right hand side accounts for third body input and the second term for third body removal. The solution of Eq. (4) is given as:

$$X(t) = \frac{K_{bc}F_cU_c + K_{br}F_rU_r}{K_{bc}F_cU_c + K_{br}F_rU_r + K_{bEc}F_cU_c + K_{bEr}F_rU_r}\left(1 - \exp\left(-\frac{t}{\tau}\right)\right). \tag{5}$$

The constants K_{bc} and K_{br} are the wear coefficients for silver between the ball-race and the ball-rod, respectively, and influence how the third body is filled with silver from incipient sliding during the test. The constants K_{bEc} and K_{bEr} are the wear coefficients for the silver that is pushed out of the wear track between the ball-race and ball-rod. The wear coefficients K_{bEc} and K_{bEr} influence how much silver is removed from the third-body due to ball sliding with the edge of the wear track during the test as shown in Figures 8 and 9. The time constant τ in equation 5 is defined as,

$$\tau = \frac{Ah_{\max}}{K_{bc}F_cU_c + K_{br}F_rU_r + K_{bEc}F_cU_c + K_{bEr}F_rU_r}, \tag{6}$$

and defines the time to steady state third-body thickness. It was found that τ also correlates with the run-in time of the RCF test configurations in Table 1. The condition $X(t) > 0$ signifies that silver is being transferred from the ball surface to the third-body volumes on the race and rod. When all of the silver has been transferred from the ball, the condition $X(t) = 1$ exists and the third body input rate goes to zero as defined in equation (4). As the third-body volume becomes depleted, that is, as silver is pushed out of the wear track as defined in the second term on the right hand side of equation (4), the test results of Figure 9 and Table 4 column 3 begins to occur. As soon as the input to the third-body volumes ceases, the volume coverage $X(t)$ diminishes resulting in asperity-to-asperity contact such that friction and vibration increase and the stopping threshold criteria of Section 2.3 is exceeded.

Equation (5) is plotted in Figure 11 using the material properties, wear coefficients, and loads presented Tables 2, 5, and 6. The wear coefficients presented in Table 5 are within the range and order of magnitude of those tested between bearing steels like Rex20 and silver, and those tested between Si_3N_4 and silver under UHV conditions found in references (Holmberg and Matthews, 2009) and (NASA/TM 1999-209088, 1999). Table 6 contains normal load and contact area calculations from the RCF test rig of Figure 2c and are used in the calculations of equations (5) and (6).

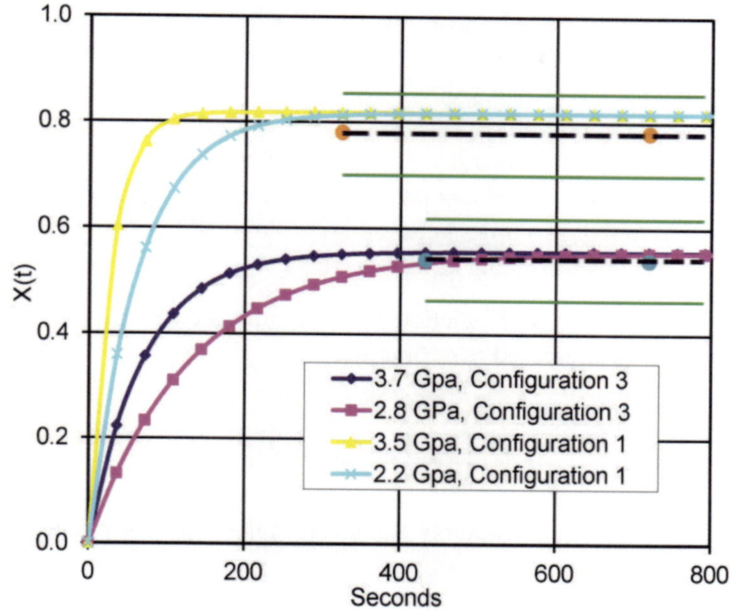

Figure 11. Fractional coverage of the third body volume calculated using equations (5) and (6) with values from Tables 2, 5, and 6 for two test configurations from Table 1.

K (m²/N)	K_{bc}	K_{br}	K_{bEc}	K_{bEr}
Test Configuration 3	1.0E-15	2.0E-16	1.0E-15	5.0E-17
Test Configuration 1	1.0E-15	2.0E-15	1.0E-15	2.0E-17

Table 5. Wear coefficients used in equations (5) and (6).

Hertzian Contact Stress (GPa)		F_r (N)	F_c (N)	Third Body Surface Area (m²)
Test Configuration 3	3.7	264	145	7.2E-05
	2.8	101	56	5.4E-05
Test Configuration 1	3.5	237	130	6.2E-05
	2.2	67	37	3.7E-05

Table 6 Normal forces and third-body contact area calculations.

Observation of the curves in Figure 11 suggests steady-state third-body coverage after 400 seconds of RCF testing. For comparison, thickness measurements of the silver remaining on the balls of suspended RCF tests reveal a third-body-coverage steady state value, X_{ss}, between 0.46 to 0.61 when testing in configuration 3 and, X_{ss} between 0.68 and 0.89 when testing in configuration 1. This data was collected from suspended tests similar to Figure 8 and do not include spall failures. These values represent a range of steady state fractional coverage for each configuration and are shown in Figure 11 as dashed lines with solid lines above and below indicating the range of measured coverage, X_{ss}.

Comparison of the measured steady state coverage, X_{ss} with that calculated from solution of equation (4) shows good agreement between measured and predicted third body fractional coverage using the wear coefficient values presented in Table 5. The trending of coverage, X_{ss}, to the same steady state values for each of the configurations 1 and 3 is due to the material type and loading conditions related to the RCF test setup. The run-in time for each of the test configurations 1 and 3 is comparable with the transient portion of the curves in Figure 11, suggesting that the volumes between asperities on the rod and races fill-up within the first 10 minutes of the test rotating at 130 Hz.

A steady state wear factor for the depletion of silver from the ball may be calculated using Archard's wear equation integrated over time as,

$$V_{ball} = \int_{0}^{t_f} K_{ball} F_{ball} U_{ball} \left(1 - X(t)\right) dt. \tag{7}$$

Solution of equation (7) and application of equation (5), the ball steady state wear factor may be expressed as,

$$\varphi = \frac{V_{ball}}{F_{ball} t_f U_{ball}} g, \tag{8}$$

where g is the gravitational constant and t_f is the time to failure based on the stopping threshold criteria 0.35g. Table 7 contains evaluation of Equation (8) using RCF depletion-failure data from Figures 6 and 7. Spall failures were not included in the wear factor calculations of Table 7. Configuration 2 shows the smallest wear factor and had the longest RCF test life. The wear factors of configurations 1 and 3 are about the same suggesting similar test-time results using either the Rex20 rod or the Si_3N_4 rod with 12.7 mm balls. The result that wear factors for configurations 1 and 3 are similar regardless of rod type suggests that most of the third-body storage volume resides on the race. This is confirmed from the autopsy results of Figures 9 and 10 in that silver has been pushed out of the wear track on the race.

Table 1 Test Configuration	Configuration 1		Configuration 2		Configuration 3	
Contact stress	3.5 GPa	2.2 GPa	4.1 GPa	3.5 GPa	3.7 GPa	2.8 GPa
Wear Factor $\left(cm^3 cm^{-1} kg^{-1}\right)$	3.47E-10	3.23E-10	7.49E-11	3.37E-11	3.12E-10	2.34E-10

Table 7. Steady state wear factor of the ball, calculated using data from all non-spall RCF tests.

4.2. Lundberg-Palmgren emperical model

Empirical modeling with ex-situ data allows coating life prediction based on past performance. RCF data collected over a range of contact stresses may be used to extrapolate coating-life within the range of the stresses tested. In this section a Lundberg-Palmgren model is used to back-calculate a basic load capacity parameter, C, for each of the test configurations of Table 1. The load capacity parameter may then be used to plan the length of any UHV-RCF test based on the test-load for each configuration. The load capacity calculation follows (Bhushan, 1999) chapter 8. The stress cycles corresponding to 10% failure maybe calculated as,

$$L_{10} = \left(\frac{C}{W}\right)^3, \tag{9}$$

where the variable W corresponds to the radial load applied to the ball, and C is the basic load capacity of the test configuration with respect to a ball-bearing type system. The basic load capacity parameter, C may be calculated using the RCF cycles-to-failure results similar to those presented in Figures 5 through 7. The values of L_{10} and W were measured for each of the test configurations shown in Table 1. Using data from Table 6 the L_{10} life for each test configuration and loading is plotted as a function of load capacity in Figure 12. The measured L_{10} life for different contact stresses is also plotted in Figure 12 as well, represented as vertical lines (large and small dashed lines, and one dash-dot line).

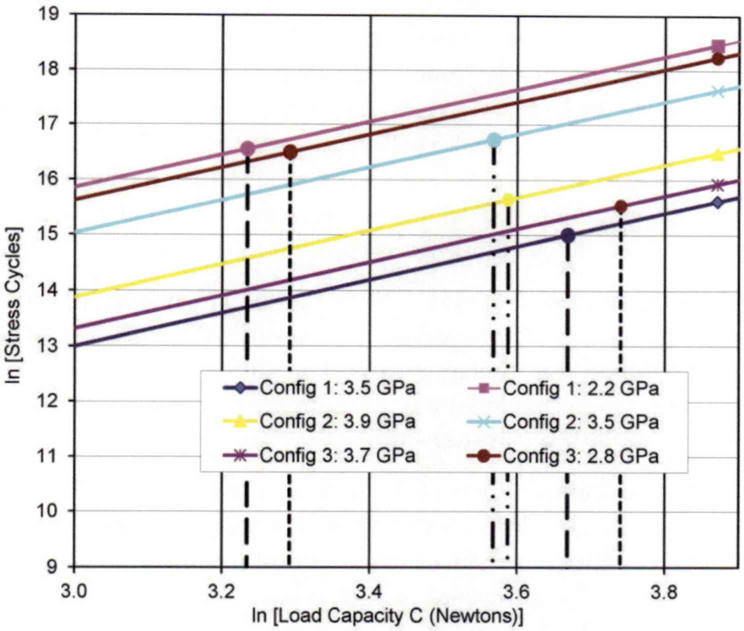

Figure 12. Plot of the natural log of L_{10} stress cycles verses load capacity parameter, C, using equation (9) for three test configurations in Table 1. Vertical lines represent measured L_{10} life for three test configurations and three loads (GPa).

The Lundberg-Palmgren model in equation (9) may require life-adjustment factors to fit the model to experimental data. Concerning Figure 12, it is to be expected that the load capacity parameter, C, will be the same for each test configuration independent of loading.

Configuration 2 demonstrates this attribute. The dash-dotted vertical lines related to configuration 2 are very near to each other, suggesting that life-adjustment factors are not needed to fit the data related to configuration 2. Equation (9) alone may be used to predict RCF life based on *W* and *C* when testing in configuration 2. In contrast, configurations 1 and 3 give different load capacity parameters for the same configurations as shown in Figure 12. The large and small dashed-lines related to Configurations 1 and 3 do not line up on the same load capacity parameter, *C*, suggesting that life-adjustment factors will be needed to accurately calculate L_{10} life for these test element combinations.

5. Conclusions

A ball-rod RCF test platform has been successfully demonstrated for testing thin solid film lubricants in ultra-high vacuum and at high rotational speeds. The film systems tested included silver and nickel-copper-silver and results indicate that film thickness initially present on the balls has the most influence over test life based on 118 tests and a thickness range of 180 to 200 nm. The Weibull shape factor for all test data was in the range of 2.5 to 2.8, indicating typical bearing-type failure modes and test-times. A shape factor in this range suggests predictability of the test method, which enables repeatable conclusions concerning tribology testing of rolling bearing elements.

For the two ball sizes investigated, the 7.94 mm ANSI T5 balls with nickel-copper-silver gave longer RCF test life than the 12.7 mm M50 steel balls for similar Hertzian contact stress loading. Surface finish of the balls and races also influenced test longevity, with longer life and higher load capability for the 7.94 mm T5 balls with a surface Ra of 0.04 compared to the 12.7 mm M50 balls with a surface Ra of 0.32. The surface roughness of the races directly affects the volume of silver that may be stored within the wear track over the course of the test. Rod material and surface finish did not influence test time for the materials Rex 20 steel and Si_3N_4.

Two types of failure modes were observed for all tests: i) early spall failure and, ii) silver film depletion. In the former, a film-spall on at least one ball resulted in increased friction and plastic yielding within the contact area leading to the onset of spall at the ball-film interface. In the latter failure mode, as the lubricant film became depleted the friction and vibration increased and the test was stopped.

Surface finish of the rod and race elements influences the rate of solid film transfer from the ball surface to the rod and races. Inspection of the wear surfaces of suspended test components suggested a third-body transfer mechanism of the lubricating film could be used to predict test longevity. A third-body self-replenishing model approach was applied and there is good agreement with measured steady state film thickness and predicted film height.

A Lundberg-Palmgren empirical model fit of the RCF test data for L_{10} life was used to calculate the load-capacity parameter for three test configurations. Stable prediction of a load capacity parameter will enable better planning for expanded used of the UHV-RCF test platform. The load capacity parameters for configurations involving 12.7 mm M50 balls with M50 races varied with applied load, suggesting that further work is needed to establish load-reduction correction factors for these configurations. Load-reduction factors for configurations involving 7.94 mm T5 balls with nickel-copper-silver film are not required. Future work with the UHV-RCF platform will involve continued testing for influence of process parameters and deposition methods, such as magnetron, on RCF test life similar to Danyluk and Dhingra (2011, 2012a). Expansion of the platform for study of surface interaction with heat and fatigue under high vacuum will also be explored.

Author details

Mike Danyluk* and Anoop Dhingra
Mechanical Engineering Department, University of Wisconsin Milwaukee, Milwaukee, Wisconsin USA

6. References

Berthier Y, Godet M, Brendle M, (1989), Velocity accommodation in friction, *Tribology Transactions*, 32: 490-496.

Bhushan B, (1999), Principles and Applications of Tribology, *John Wiley & Sons, New York*, Chapters 4 and 8.

Danyluk M, Dhingra A, (2011), Rolling Contact Fatigue in High Vacuum Using Ion Plated Nickel-Copper-Silver Solid Lubrication, *J. Vac. Sci. Technol. A*, 29: 011005.

Danyluk M, Dhingra A, (2012a), Rolling contact fatigue using solid thin film lubrication, *Wear* 274-275: 368 - 376.

Danyluk M, Dhingra A, (2012b), Influence of process parameters on rolling-contact-fatigue life of ion plated nickel- copper-silver lubrication, *J. Vac. Sci. Technol. A*, 30: 031502.

Higgs C, Wornyoh E., (2008), An in situ mechanism for self-replenishing powder transfer films: Experiments and modeling, *Wear* 264: 131-138.

Holmberg K, Matthews K (2009) Coatings Tribology Properties, Mechanisms, Techniques and Applications in Surface Engineering, Volume 10: 202-208, Elsevier.

Hoo J, (1982), A Ball-Rod Rolling Contact Fatigue Tester, *ASTM STP 771 ASTM* pp.107-124.

Hu Y, Zhu D, (2000), A Full Numerical Solution to the Mixed Lubrication in Point Contacts, *Journal of Tribology*, Vol. 122, Issue 1, pp.1-9.

Matthews A, Franklin S, Holmberg K, (2007), Tribological coatings: contact mechanisms and selection, *J. Phys. D: Appl. Phys* 40: 5463-5475.

NASA report SP-5059(01), (1972), Solid Lubricants: A Survey, Technology Utilization Office NASA.

NASA/TM 1999-209088/Part1, (1999), Friction and Wear Properties of Selected Solid Lubricating Films, Part 1: Bonded and Magnetron-Sputtered Molybdenum Disufide and Ion-Plated Silver Films.

Polonsky I, Chang T, Keer L, Sproul W, (1998), A Study of rolling contact fatigue of bearing steel coated with PVD TiN films: Coating response to cyclic contact stress and physical mechanisms underlying coating effect on the fatigue life, *Wear* 215: 191-204.

* Corresponding Author

Rosado L, Forster N, Thompson K, Cooke J, (2010), Rolling Contact Fatigue Life and Spall Propagation of AISI M50, M50NiL, and AISI 52100, Part 1: Experimental Results, *Tribology Transactions*, 53: 29-41.

Sadeghi F, Jalalahmadi B, Slack T, Raje N, Arakere N, (2009), A Review of Rolling Contact Fatigue, *Journal of Tribology*, Vol. 131, Issue 4, pp.1-15.

Wornyoh E, Higgs C, (2011), An asperity-based fractional coverage model for transfer films on a tribological surface, *Wear* 270: 127-139.

Chapter 5

Bearing Fault Diagnosis Using Information Fusion and Intelligent Algorithms

Jiangtao Huang

Additional information is available at the end of the chapter

http://dx.doi.org/10.5772/51079

1. Introduction

Rotating machinery is very common in industrial systems, and it plays an important role in industrial development and economic development. With the rapid advancement in industry, rotating machinery is becoming more and more complex and require constant attention. Although the reliability and robustness of rotating machinery also have been improving, some occasional failure events of components often lead to unexpected downtime while resulting in huge losses. And rolling element bearing is often at the heart of these rotating machinery which suffers from fault more frequently. These faults may cause the machine to break down and decrease its level of performance [6]. So, it is urgent to diagnose the incipient errors exactly in these bearings.

In traditional fault diagnosis, a single sensor is always used to get the operation conditions of several machine components. The collected signal involves many correlated features [33]. During operating process, the machine set can generate many kinds of signals. And those approaches based on the vibration signal analysis are advantageous because of their visual feature, easy measurability, high accuracy and reliability [34]. Fault diagnosis using raw vibration signals, a wide variety of techniques have been introduced in recent years. There are mainly including signal processing methods and intelligent systems application. Signal processing methods are traditional methods which are still common used, such as wavelet and wavelet packet methods [23–25], empirical mode decomposition [15, 35], time-frequency distributions [7], blind source separation [29]. While intelligent system approaches for fault diagnosis are including artificial neural networks (ANNs) [36], support vector machines (SVMs) [33], adaptive neuro-fuzzy inference system (ANFIS) [19] and fuzzy technique [28], etc.. These approaches are based on one data source or individual decision system, and many researchers have realized and shown that an individual decision system with a single data source can only acquire a limited classification capability which may not be enough for a particular application [22]. So, it is necessary to combine multiple decision systems to carry on failure diagnosis.

Multi-sensor information fusion is an emerging interdisciplinary beginning in the military field, and it has already been successfully applied in many different areas. In the field of industrial equipment fault diagnosis, multi-source information fusion technology application is still in its early stage. Multi-sensor information fusion is divided into three levels: sensor level, feature level and decision level. And multiple classifier ensemble approach belongs to decision level information fusion. In the recent years, the use of multiple classifiers has gained a lot of attention and researches have continuously showed the benefits of using multiple classifiers to solve complex problems [4]. In contrast, the feature-level fusion has not probably received the amount of attention it deserves [32].

By using information fusion theory, this chapter will introduce some bearing fault diagnosis approaches. And these methods can divide into two categories: fault diagnosis based on feature-level fusion [11] and fault diagnosis based on decision-level fusion [14]. In the proposed fusion methods for bearing fault diagnosis, some intelligent algorithms are used for feature dimension reduction or pattern recognition. The feature-level fusion approach for bearing fault diagnosis is using gene expression programming (GEP), while the decision-level fusion approach using multiple classifier ensemble method. And the decision-level fusion approach is based on the new bearing fault diagnosis method [12] which uses empirical mode decomposition (EMD) and fractal feature parameter classification.

2. Bearing fault diagnosis using fractal feature parameter classification

Faulty and normal machine conditions are always treated as classification problems based on learning pattern from empirical data modeling in complex mechanical processes and systems [31]. In this approach, a general framework for applying classification methods to fault diagnosis problems includes two steps: representative feature extraction and pattern classification. Feature extraction is a mapping process from the measured signal space to the feature space. Representative features which demonstrate the information of fault are extracted from the feature space. Pattern classification is the process of classifying the extracted features into different categories by geometric, statistic, neural or fuzzy classifiers. And recently, the development of artificial intelligence techniques has led to their application in fault diagnosis area. Meanwhile, artificial neural networks (ANNs) and support vector machines (SVMs) have been successfully applied to the intelligent fault diagnosis of mechanical equipment [27].

In practice, the classical approach is not always reliable when the extracted features are contaminated by noise. And most intelligent fault diagnosis approaches are complex, especially in solving multiple fault diagnosis problems. In this section, a novel, simple, fast and reliable intelligent method for solving multiple fault diagnosis problem will be proposed. And this approach is based on EMD and fractal feature parameter extraction.

2.1. Methodology

Fractal dimension is considered right from its invention [21] to be a good parameter to characterize time sequences of values of natural variables. And a simple, fast and accurate method for calculating the fractal dimension of data's time sequences was presented by Sy-Sang liaw and Feng-Yuan Chiu [20]. This method considers that a time sequence of $2^M + 1$ values is separated by a constant time interval which is well fitted by a fractal function $f(t)$ in the period $[0, T]$. Then, calculating the fractal dimension D of $f(t)$ by using the known values of $f(t)$ at $t_j = jT/2^M$. To achieve this aim, Liaw and Chiu first defined $L_k(f)$, the piecewise linear interpolation of level $k(k = 0, 1, 2, ..., M)$, to $f(t)$ as the union of the line segments connecting the points $[t_j, f(t_j)]$ and $[t_{j+1}, f(t_{j+1})]$, where $t_j = jT/2^k$, $j = 0, 1, 2, ..., 2^k$ (see Figure 1). And then they checked out how poor the interpolation function $L_k(f)$ is relative to the next level of interpolation $L_{k+1}(f)$. The error of $L_k(f)$ is defined as the sum of the absolute value of the differences of $L_k(f)$ and $L_{k+1}(f)$ at all $t_j = jT/2^{k+1} \equiv j\varepsilon_k$:

$$\Delta_k \equiv \sum_{j=0}^{2^{k+1}} |L_{k+1}(f(t_j)) - L_k(f(t_j))| = \sum_{j=odd}^{2^{k+1}} |f(t_j) - \frac{f(t_j - \varepsilon_k) + f(t_j + \varepsilon_k)}{2}|, \ t_j = j\varepsilon_k \quad (1)$$

Liaw and Chiu [20] found that the value Δ_k is proportional to $(\varepsilon_k)^{1-D}$ when k is large enough.

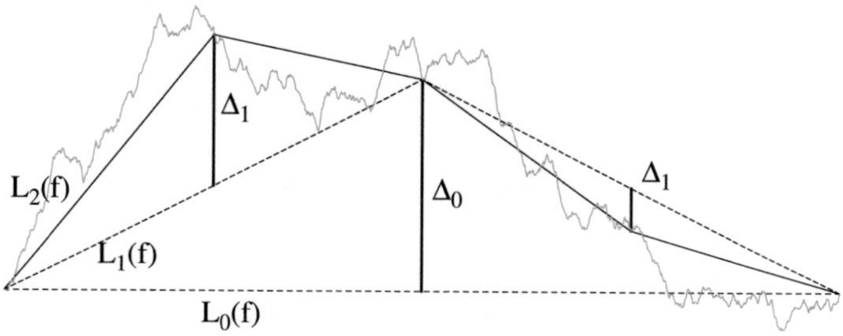

Figure 1. Piecewise interpolation $L_k(f)$ to a function $f(t)$ (grey) at level 0 (dotted), 1 (dashed), and 2 (solid). Δ_k (thick solid) denotes the error of the kth level interpolation with respect to the $k+1$ level [20]

Thus, the fractal dimension D of $f(t)$ can be obtained from the slope s of the log-plot of Δ_k with respect to the level k by $D = 1 + s/log2$ for large enough k values.

In this bearing fault diagnosis method, raw vibration signal will be seen as a time sequences of data. Raw vibration signal is often heavily clouded by various noises due to the compounded effect of other machine elements' interferences and background noises presenting in the measuring device [2]. So, EMD is used to analysis raw vibration signal to filter noise before extracting its fractal feature. As discussed by Huang et al. [10], the EMD method is designated to deal with non-stationary and nonlinear signals. This method is based on the simple assumption that any data consists of different simple intrinsic modes of oscillations. Using the EMD method, complicated signals can be decomposed in a finite set of intrinsic mode functions (IMFs). Each IMF should meet the following two conditions: (1) in the whole data set of a signal, the number of extreme and the number of zero crossings must either equal or differ at most by one, and (2) at any time point, the mean value of the envelope defined by the local maxima and the envelope defined by the local minima is zero.

Assume $x(t)$ is a vibration signal, and its empirical mode decomposition process can be described by following steps:

Step 1 Initialize: $r_0(t) = x(t), i = 1$.

Step 2 Extract the i-th intrinsic mode function (IMF) $c_i(t)$:

Step 2.1. Initialize: $h_0(t) = r_{i-1}(t), j = 1$.

Step 2.2. Determine all the maximal values, minimal value points of $h_{j-1}(t)$ and fit all extreme points into the upper and lower envelope of the original signal with the cubic spline line.

Step 2.3. Determine the mean value of the upper and lower envelope of $h_{j-1}(t)$, designated as $m_{j-1}(t)$.

Step 2.4. Calculate the difference between $h_{j-1}(t)$ and $m_{j-1}(t)$, $h_j(t)$: $h_j(t) = h_{j-1}(t) - m_{j-1}(t)$.

Step 2.5. If $h_j(t)$ satisfies the conditions of IMF, then it is designated as $c_i(t) = h_j(t)$. Otherwise, update the value of j: $j = j + 1$, and return to Step 2.2.

Step 3. Get the remaining signal: $r_i(t) = r_{i-1}(t) - c_i(t)$, after decomposing the i-th IMF.

Step 4. When $c_i(t)$ or $r_i(t)$ satisfies the given termination condition, the cycle is ended. Designate the final remaining signal as $r_n(t)$ $(n = i)$. Otherwise, update the value of i: $i = i + 1$, and return to Step 2.

Finally, raw vibration signal can be decomposed into n IMFs: $c_i(t), i = 1, ..., n$ and one residue function $r_n(t)$:

$$x(t) = \sum_{i=1}^{n} c_i(t) + r_n(t) \tag{2}$$

In this work, representative feature is fractal feature parameter extracting from each IMF. Because the method of fractal dimensions of time sequences needs k to be large enough, we use fractal feature parameter. And fractal feature parameter of each IMF will be calculated as Equation 3 shows. It is easy to know that the IMF's numbers of different raw vibration signal samples are different. And in the vibration signal examination, we find that the rich operating condition information is inside the front IMFs. So, we can integrate the residual IMFs into a component. In this new method, a parameter L is set to denote the number of IMF using to extract representative feature. And the L-th IMF will be re-denoted as $c_r(t)$ whose calculation form as Equation 4. Then, the feature set of each raw signal has L fractal feature parameters. For example, we set the value of parameter L as $L = 6$. Figure 2 summarizes all the IMFs and fractal features obtained from a bearing inner race fault signal sample. Table 1 presents the fractal feature parameters of IMFs of different operating condition vibration signal samples. And from Table 1, it is clear that fractal feature parameter sets of the same operating condition are similar, and it is easy to distinguish different operating conditions of fractal feature parameter.

$$p = \sum_{k=0}^{M-1} \Delta_k \tag{3}$$

$$c_r(t) = \sum_{i=L}^{n} c_i(t) \tag{4}$$

2.2. Results and discussion

By using fractal feature parameter classification, bearing fault diagnosis method is applied to the bearing fault signal analysis from the Case Western Reserve University website [3]. The ball bearings are installed in a motor driven mechanical system, as shown in Figure 3. By a self-aligning coupling, a three-phase induction motor is connected to a dynamometer

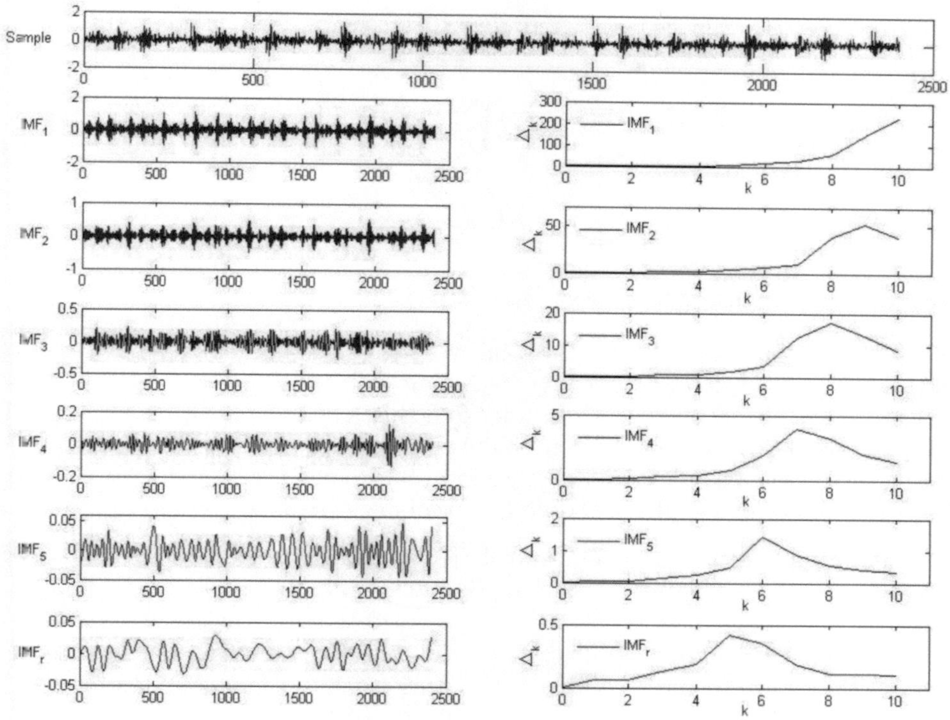

Figure 2. The resulting empirical mode decomposition components and fractal features from the inner race fault signal sample

Operating condition	Sample index	$p(c_1)$	$p(c_2)$	$p(c_3)$	$p(c_4)$	$p(c_5)$	$p(c_r)$
Normal	1	62.5909	26.8249	8.9951	6.6883	1.7284	0.6491
	2	67.8337	25.7203	8.9875	5.9469	2.8435	0.4777
Outer race	1	442.1013	45.3849	17.8250	5.9114	3.0012	0.6935
	2	479.2442	44.4325	19.1783	6.6492	3.4410	0.4973
Inner race	1	487.9925	151.8692	59.1485	14.1692	4.6696	1.5808
	2	511.5880	149.7561	68.9864	28.0217	5.8445	2.6485
Ball	1	295.3898	32.5473	19.6932	4.1315	2.0519	0.3888
	2	270.6188	33.8913	20.9217	4.4377	2.3469	0.3063

Table 1. Fractal feature parameters of different operating condition samples (defect size: 0.007inches)

and a torque sensor. The bearings are installed in a motor driven mechanical system. The dynamometer is under control so that desired torque load levels can be achieved. Vibration data is collected using accelerometer, which is attached to the housing with magnetic bases. Accelerometer is placed at the 12 o'clock position at the driven end of the motor housing. In machine condition monitoring, an accelerometer can provide rich information about conditions of several machine components. For example, the measured data from the accelerometer in this experiment is a mixture of signals reflecting conditions of the bearing inner race, outer race and rolling elements. The vibration data are collected by a 16 channel DAT recorder with 12,000 Hz.

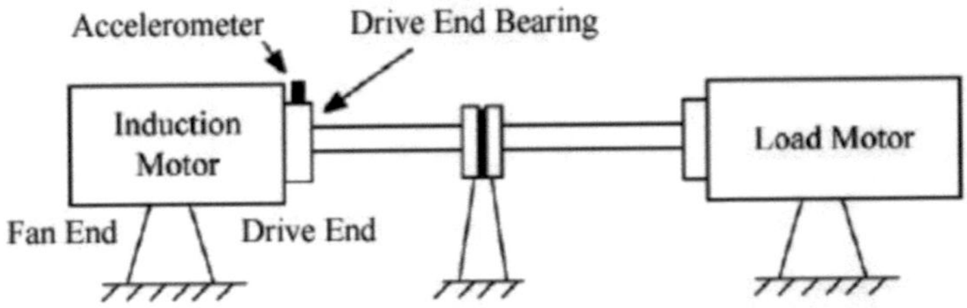

Figure 3. Schematic diagram of the experimental setup

As we referred in Figure 3, in the mechanical system, single point faults were introduced to the test bearings using electro-discharge machining with fault diameters of 7 mils, 14 mils and 21 mils. Each bearing was tested under four different loads (0, 1, 2 and 3 hp). Three bearing data sets (A-C) were obtained from the experimental system under the four different operating conditions: (1) under normal condition, (2) with outer race fault, (3) with inner race fault, (4) with ball fault. The detailed descriptions of the three data sets are shown in Table 2.

Data set	The number of training sample	The number of testing sample	Defect size(inches) (training/testing)	Operating condition	Class label
A	40	40	0/0	Normal	1
	40	40	0.007/0.007	Outer race fault	2
	40	40	0.007/0.007	Inner race fault	3
	40	40	0.007/0.007	Ball fault	4
B	40	40	0/0	Normal	1
	40	40	0.007/0.021	Outer race fault	2
	40	40	0.007/0.021	Inner race fault	3
	40	40	0.007/0.021	Ball fault	4
C	40	40	0/0	Normal	1
	40	40	0.007/0.007	Outer race fault	2
	40	40	0.007/0.007	Inner race fault	3
	40	40	0.007/0.007	Ball fault	4
	40	40	0.014/0.014	Outer race fault	5
	40	40	0.014/0.014	Inner race fault	6
	40	40	0.014/0.014	Ball fault	7
	40	40	0.021/0.021	Outer race fault	8
	40	40	0.021/0.021	Inner race fault	9
	40	40	0.021/0.021	Ball fault	10

Table 2. Description of three data sets

Data set A is formed by 320 samples. These samples include 4 different operating information under 4 conditions (0, 1, 2 and 3 hp), and among which the fault defect size is 0.007 inches. Every operating condition has 80 data samples. The whole data set is divided into 2 parts: 160 samples for training and 160 for testing. So, the task can be viewed as a four-class classification aimed at 4 different operating conditions. Data set B also contains 320 data samples. The training samples are including samples with 0.007 inches fault defect, while testing samples 0.021 inches fault defect. Data set B is used to further investigate the performance of fault diagnosis scheme. Data set C comprises 800 data samples including three different defect sizes of 0.007, 0.014 and 0.021 inches under four different loads. It covers four different operating conditions, too. Each class data subset has been partitioned into two equal halves, one partition is used for training, while the other for testing. The purpose of data set C is to test the reliability of the novel approach in identifying the various grades of fault.

In order to evaluate the classification performance of the fractal feature parameter of IMF, orthogonal quadratic discriminant function (OQDF-E) [9] is used to train and test on three data sets showed in Table 2. Table 3 gives the classification performance on various data sets. The new bearing fault diagnosis method can get good decision accuracy as Table 3 shows. Table 4 extends the analysis of results and shows the classification performance between normal and fault operating condition. From Table 4, we can see that the new method using fractal feature parameter can get perfect performance in fault detection.

Data set	A	B	C
Train accuracy (%)	100	100	78.25
Test accuracy (%)	100	78.13	80

Table 3. Classification performance

	Data set A		Data set B		Data set C	
Operating condition	Normal	Fault	Normal	Fault	Normal	Fault
Test accuracy (%)	100	100	100	100	100	100

Table 4. Fault detection performance

3. Decision-level fusion for bearing fault diagnosis

In above section, we have proposed a simple, fast and good performance fault diagnosis approach. This approach is based on single sensor source and using individual classifier. It can obtain high accuracy on the multiple fault types recognition problems under the same fault degree. But when under multiple fault degrees, it declines in performance. To deal with this problem, this section will introduce a new method based on decision-level fusion for bearing fault diagnosis. The new fusion method includes four stages. These four stages are vibration signal acquisition and decomposition, fractal feature parameter extraction, single data source fault diagnosis and decision-level fusion for fault diagnosis. The first three stages are the same with the method described in the above section. So, we only state the last step in this section.

3.1. Methodology

Given a specific pattern recognition problem, different classifier has different classification performance. Very satisfactory results can not always be got if we simply conduct a study on a single classifier to improve its classification accuracy. Multiple classifier system (MCS) can overcome limitations of individual classifier and enhance classification accuracy. The techniques of combining the outputs of several classifiers have been applied to a wide range of real problems and it has been shown that MCSs outperform the traditional approach of using a single high-performance classifier [26].

The most often used classifiers combination approaches in MCS include the majority voting [30], the weighted combination (weighted averaging) [18], the probabilistic schemes [16, 17], the Bayesian approach (naïve Bayes combination) [1, 18, 30], the Dempster-Shafer (D-S) theory of evidence [5, 30] and etc. This section will propose a new classifiers combination method which treats the combination process as linear programming problem.

Assume that K base classifiers are used in MCS, and M kinds of fault states including normal condition on the bearing fault diagnosis problem. Then, a decision matrix can be given as follow in the process of multiple classifiers combination.

$$D(x) = \begin{bmatrix} P_1(F_1|x) & P_1(F_2|x) & \dots & P_1(F_M|x) \\ P_2(F_1|x) & P_2(F_2|x) & \dots & P_2(F_M|x) \\ \vdots & \vdots & \ddots & \vdots \\ P_K(F_1|x) & P_K(F_2|x) & \dots & P_K(F_M|x) \end{bmatrix} \quad (5)$$

The new method introduced in this section will fuse those posterior probabilities in the decision matrix for constructing a global classifier E to make final decision. The posterior probability output of global classifier E for each fault state is calculated by following mode:

$$P_E(F_i|x) = \sum_{k=1}^{K} \beta_k P_k(F_i|x), \quad \forall\, i \in \{1, 2, \dots, M\}, \quad (6)$$

where β_k ($\sum_{k=1}^{K} \beta_k = 1$) is a dynamic association weight in MCS.

This new decision-level fusion method for bearing false diagnosis is based on the assumption: the base classifier has higher real-time recognition accuracy, if its posterior probabilities of all fault states are greater difference. That is to say, if individual decision system very determines that current operating condition belongs to a certain type of fault states, the posterior probability of the certain fault state will much higher than others. Using this hypothesis, the problem of multiple classifiers combination can be converted into a linear programming problem. And the objective function of this linear programming is defined as: $\sqrt{[\sum_{i=1}^{M} (P_E(F_i|x) - 1/M)^2]/M}$.

In current using classifier ensemble methods, base classifier's statistical performance is a major consideration factor. But we find the realtime decision information also can be a consideration factor. And in the new MCS method, we use within-class decision support [13] which is defined as: within-class decision support indicates that base classifier individual class recognition output gets the decision support degree from other same class recognition outputs in MCS. This decision support degree is measured by the difference between current output and its nearest output. For example, the within-class decision support of $P_k(F_i|x)$ which denotes posterior probability of the i-th state from the k-th base classifier is: $1 - \min_{1 \le k' \le K, k' \ne k} |P_k(F_i|x) - P_{k'}(F_i|x)|$.

Real-time decision support value (DSV) of base classifier in MCS is the sum of all class recognition output's within-class decision support value. And it is easy to get its calculation formula as:

$$\mu_k = \sum_{i=1}^{M} (1 - \min_{1 \le k' \le K, k' \ne k} |P_k(F_i|x) - P_{k'}(F_i|x)|), \quad \forall\, k \in \{1, 2, \dots, K\} \quad (7)$$

In the proposed decision-level fusion method for bearing fault diagnosis, we set a rule: if the real-time decision support value of base classifier is higher, its dynamic association weight of it is bigger. And this rule can be described as follows:

$$if\ \ \mu_k < \mu_{k'}\ \ then\ \ \beta_k < \beta_{k'},\ \ \forall\, k,k' \in \{1,2,\ldots,K\}, k' \neq k \tag{8}$$

$$if\ \ \mu_k = \mu_{k'}\ \ then\ \ \beta_k = \beta_{k'},\ \ \forall\, k,k' \in \{1,2,\ldots,K\}, k' \neq k \tag{9}$$

$$if\ \ \mu_k > \mu_{k'}\ \ then\ \ \beta_k > \beta_{k'},\ \ \forall\, k,k' \in \{1,2,\ldots,K\}, k' \neq k \tag{10}$$

That is to say, the relationship between dynamic association weights is determined by the relationship between real-time decision support values of different base classifiers. And these relationships will be used in the linear programming problem by the form of relationship vectors. Relationship vectors are defined as Table 5 shows (for example, $K = 3$). From Table 5, it is clear that each real-time relationship between DSVs is re-expressed by one or two relationship vectors. And it is also clear that each relationship vector is K dimensions. All these relationship vectors compose a relationship matrix which is denoted as R.

Real-time relationship between DSVs	Relationship vectors
$\mu_1 < \mu_2$	$[1\ \ -1\ \ 0]$
$\mu_1 = \mu_2$	$[-1\ \ 1\ \ 0]$ & $[1\ \ -1\ \ 0]$
$\mu_1 > \mu_2$	$[-1\ \ 1\ \ 0]$

Table 5. Example for relationship vector construction (K=3)

In order to simplify the fusion formula, a $K \times 1$ matrix β ($\beta = [\beta_1; \beta_2; \ldots; \beta_K]$) is used to replace β_k ($k \in \{1,2,\ldots,K\}$) for calculating the decision output of global classifier E. Then, Equation 6 can be transformed into following simplified form: $P_E(x) = D(x)^T\beta$. And the objective function $\sqrt{[\sum_{i=1}^{M}(P_E(F_i|x) - 1/M)^2]/M}$ is simplified as: $\sqrt{\sum_{i=1}^{M}(P_E(F_i|x) - 1/M)^2}$, then further simplified to: $||D(x)^T\beta - \frac{1}{M}||_2$. Finally, use relationship matrix R to formulize constraint rules by the form: $R\beta \leq 0$. Now, we can give complete linear programming problem description as Equation 11 shows, where N is the count of relationship vectors of current relationship matrix.

$$\begin{aligned} \max \quad & ||D(x)^T\beta - \frac{1}{M}||_2 \\ subject\ to \quad & [1]_{1\times K}\beta = 1 \\ & [0]_{K\times 1} \leq \beta \leq [1]_{K\times 1} \\ & R\beta \leq [0]_{N\times 1} \end{aligned} \tag{11}$$

Solving the linear programming problem as above, we can obtain the dynamic association weight matrix β. Using this dynamic association weight matrix, the fusion decision vector of global classifier E can be calculated. And the final decision of bearing fault diagnosis can be got by:

$$E(x) = i\ \ with\ \ P_E(F_i|x) = \max_{1\leq i'\leq M} P_E(F_{i'}|x) \tag{12}$$

3.2. Results and discussion

The decision-level fusion method for bearing fault diagnosis is also applied to the rotating machinery from the Case Western Reserve University website [3]. In this experiment, vibration signals are collected from accelerometers which attached to the motor at different positions as Figure 4 shows. And dynamometer is used to control the torque load level. In this work, we study four different operating conditions recognition under four different loads (0, 1, 2 and 3 hp) with fault diameters of 7 mils, 14 mils and 21 mils. And these four operating conditions are normal condition, outer race fault, inner race fault and ball fault.

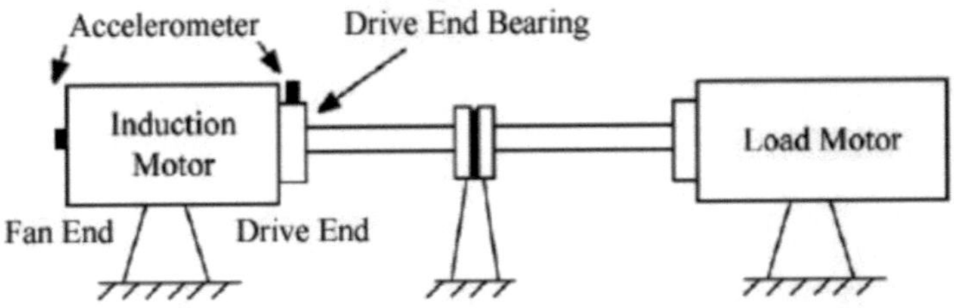

Figure 4. Schematic diagram of the experimental setup

Two data sets are constructed as Table 6 presents for testing the diagnosis performance of new decision-level fusion method. Each data set samples cover four different operating conditions and four different loads. And each class of two data sets has 160 data samples which are divided into two equal halves, one for training and the other for testing. Data set A is a four-class classification task corresponding to the four operating conditions. Data set B is a ten-class classification task corresponding to various grades of different faults.

Data set	The number of training sample	The number of testing sample	Defect size(inches) (training/testing)	Operating condition	Class label
A	80	80	0/0	Normal	1
	80	80	0.007/0.007	Outer race fault	2
	80	80	0.007/0.007	Inner race fault	3
	80	80	0.007/0.007	Ball fault	4
B	80	80	0/0	Normal	1
	80	80	0.007/0.007	Outer race fault	2
	80	80	0.007/0.007	Inner race fault	3
	80	80	0.007/0.007	Ball fault	4
	80	80	0.014/0.014	Outer race fault	5
	80	80	0.014/0.014	Inner race fault	6
	80	80	0.014/0.014	Ball fault	7
	80	80	0.021/0.021	Outer race fault	8
	80	80	0.021/0.021	Inner race fault	9
	80	80	0.021/0.021	Ball fault	10

Table 6. Description of two data sets

These data samples are extracted from two different sensor sources. And the number of samples from each sensor source is half of the total. If each sensor source's samples are seen as a subset of data set, each data set has two subsets. For example, data set A has two subsets:

A1 and A2. A1 is composed by the samples from the driven end accelerometer, while A2 from the fan end accelerometer. Table 7 gives the elements description of data set A in detail.

Data set	Class label	Sub dataset	Sensor source	The number of training sample	The number of testing sample
A	1	A1	driven end accelerometer	40	40
		A2	fan end accelerometer	40	40
	2	A1	driven end accelerometer	40	40
		A2	fan end accelerometer	40	40
	3	A1	driven end accelerometer	40	40
		A2	fan end accelerometer	40	40
	4	A1	driven end accelerometer	40	40
		A2	fan end accelerometer	40	40

Table 7. The elements description of data set A

In this work, two different classifiers, k-NN ($k = 7$) and Parzen classifier, are used for fault diagnosis task. And these two different classifiers identify rotating machinery operating condition using vibration signals collected from driven end and fan end accelerometers respectively. That is to say, each data set has four individual decision system results. And MCS is composed by these four base classifiers.

Table 8 gives individual classifier recognition accuracy on subsets of data set A and B. It is clear that individual classifiers can attain high bearing diagnosis accuracy on data set A, but they can not maintain the same high-performance on data set B whose fault diagnosis task is extended to various grades of different fault conditions.

Data set	Subset	Classifier	Training accuracy	Testing accuracy
A	A1	k-NN classifier	100%	100%
		Parzen classifier	100%	98.75%
	A2	k-NN classifier	100%	100%
		Parzen classifier	100%	99.38%
B	B1	k-NN classifier	88%	87.5%
		Parzen classifier	100%	85.75%
	B2	k-NN classifier	91.75%	88%
		Parzen classifier	100%	86.75%

Table 8. Fault diagnosis performance using base classifier

Table 9 shows the fault diagnosis performance of the novel decision-level fusion model using multiple classifier system. It is clear that the novel decision-level fusion model can get high recognition accuracy even in the difficult fault diagnosis task. In the testing phase of data set B, fault diagnosis accuracy of the new fusion model is higher than all base classifiers' accuracy as Table 8 shows. And it increases 6.5 percentage points averagely.

Data set	Training accuracy	Testing accuracy
A	100%	100%
B	100%	93.5%

Table 9. Fault diagnosis performance using the new fusion model

To further analyze performance of the new fusion model, a new k-NN classifier ($k = 3$) is added to multiple classifier system. The new multiple classifier system is used to test fault diagnosis performance on data set B. And sum rule is used to compare with the new approach. The comparison results are presented in Table 10. From Table 10, it is clear that the new approach attains the highest diagnosis accuracy.

7-NN classifier on B1	7-NN classifier on B2	Parzen classifier on B1	Parzen classifier on B2	3-NN classifier on B1	3-NN classifier on B2	Sum rule	New method
87.5%	88%	85.75%	86.75%	90%	87.75%	94.75%	95%

Table 10. Further comparison results of fault diagnosis performance

4. Feature-level fusion for bearing fault diagnosis

This section will propose a new multiple sources feature-level fusion model for bearing fault diagnosis using GEP. At present, the research of fault diagnosis based on feature-level fusion is still less, far from decision-level fusion attention. This is mainly because feature-level fusion is more difficult. But feature-level fusion application for fault diagnosis can be more effective to extract fault feature information. It is a way to improve the performance and robustness of bearing fault diagnosis system.

4.1. Methodology

GEP was invented by Ferreira [8], and it is the natural development of genetic algorithms and genetic programming. GEP uses linear chromosome which is composed of genes containing terminal and non-terminal symbols. Chromosomes can be modified by mutation, transposition, root-transposition, gene transposition, gene recombination, one-point and two-point recombination. GEP genes are composed of a head and a tail. The head contains function (non-terminal) and terminal symbols, while the tail contains only terminal symbols. For each problem, the head length (denoted h) is chosen by users, and then the head length is used to evaluate the tail length (denoted t) by: $t = (n - 1) \times h + 1$, where n is the number of arguments of the function with most arguments.

The flow of GEP is as follows:

Step 1. To set control parameters, select function classes, initialize population.

Step 2. To parse chromosome, evaluate population.

Step 3. To take use some operation such as selection, mutation, inserts sequence, recombine, mutation of random constant and inserts sequence of random constant to create new population.

Step 4. To implement best preservation strategy.

Step 5. If obtain most precision of computing, evolution would be finished, else turn to Step 2.

The new feature-level fusion model using GEP will be dealt with multiple sensors fusion problem. Assume that there are I sensors used in machine condition monitoring. For each sensor, the raw signal is divided into some signals by the same time segment. Each of these signals is processed to extract some features. In this chapter, machine operating signal features only take into account the time-domain statistical characteristics. These feature parameters of time-domain are presented in Eequations. (13-23), where $x(t)$ is a signal series and N is its number of data points.

$$p_1 = \frac{1}{N}\sum_{n=1}^{N} x(n) \tag{13}$$

$$p_2 = \sqrt{\frac{\sum_{n=1}^{N}(x(n) - p_1)^2}{N-1}} \tag{14}$$

$$p_3 = \sqrt{\frac{\sum_{n=1}^{N} x(n)^2}{N}} \tag{15}$$

$$p_4 = (\frac{\sum_{n=1}^{N}\sqrt{|x(n)|}}{N})^2 \tag{16}$$

$$p_5 = \max|x(t)| \tag{17}$$

$$p_6 = \frac{\sum_{n=1}^{N}(x(n) - p_1)^3}{(N-1)p_2^3} \tag{18}$$

$$p_7 = \frac{\sum_{n=1}^{N}(x(n) - p_1)^4}{(N-1)p_2^4} \tag{19}$$

$$p_8 = \frac{p_5}{p_3} \tag{20}$$

$$p_9 = \frac{p_5}{p_4} \tag{21}$$

$$p_{10} = \frac{p_3}{\frac{1}{N}\sum_{n=1}^{N}|x(n)|} \tag{22}$$

$$p_{11} = \frac{p_5}{\frac{1}{N}\sum_{n=1}^{N}|x(n)|} \tag{23}$$

In the pattern recognition process of bearing fault diagnosis, we assume that there are M conditions including normal condition. Let S_m^i represents the set of all training samples belonging to m-th condition ($1 \leq m \leq M$) from the i-th sensor source. Feature-level fusion model is seek a way to fuse these features from different sensor sources. The new feature-level fusion model using GEP fuses these features by looking for a feature recognition function φ which maps the feature space to another space where samples in the same class are similarity and samples dissimilarity otherwise. And then, the feature recognition function φ will direct the building of a multi-source feature fusion model in reverse direction.

Functions $+, -, \times, /, sqrt, exp$ are selected as input functions of GEP. The generation is set 5000, and fitness function is defined as:

$$Fitness = \frac{\sum_{m=1}^{M-1}\sum_{m'=m+1}^{M}(\sigma_m - \sigma_{m'})^2}{\sum_{m=1}^{M}\sum_{i=1}^{I}\sum_{k\in S_m^i}(\varphi(P_k^i) - \sigma_m)^2} \tag{24}$$

where σ_m is the mean of all m-th condition samples function mapping values, its formula is:

$$\sigma_m = \frac{1}{I}\sum_{i=1}^{I}\frac{\sum_{k\in S_m^i}\varphi(P_k^i)}{|S_m^i|} \tag{25}$$

After GEP training, a perfect feature recognition function φ can be got. Using function φ, we can calculate the mean mapping value of each operating condition samples from a certain sensor source. For building the multi-source feature evaluation matrix, the samples which are correctly classified are selected to calculate their mean. Multi-source feature evaluation matrix is composed by these mean values as Equation 26 shows.

$$\begin{bmatrix} \rho_1(1) & \rho_1(2) & \cdots & \rho_1(11) \\ \rho_2(1) & \rho_2(2) & \cdots & \rho_2(11) \\ \vdots & \vdots & \ddots & \vdots \\ \rho_M(1) & \rho_M(2) & \cdots & \rho_M(11) \end{bmatrix} \tag{26}$$

In Equation 26, each element represents the mean value of each feature component of each operating condition. For example, $\rho_2(1)$ represents the mean of all correctly classified samples of the first feature from the 2-th operating condition.

4.2. Results and discussion

In order to evaluate the proposed feature-level fusion model, we apply it to bearing fault diagnosis. And data of this bearing fault diagnosis task are also take from a lab of the Case Western Reserve University website [3]. In this work, three experiments over three data sets are conducted as Table 11 shows. Those data are collected under various operating loads from motor driven end and fan end accelerometers.

Data set	The number of training sample	The number of testing sample	Defect size(inches) (training/testing)	Operating condition	Class label
A	80	80	0/0	Normal	1
	80	80	0.007/0.007	Outer race fault	2
	80	80	0.007/0.007	Inner race fault	3
	80	80	0.007/0.007	Ball fault	4
B	80	80	0/0	Normal	1
	80	80	0.007/0.021	Outer race fault	2
	80	80	0.007/0.021	Inner race fault	3
	80	80	0.007/0.021	Ball fault	4
C	80	80	0/0	Normal	1
	80	80	0.021/0.007	Outer race fault	2
	80	80	0.021/0.007	Inner race fault	3
	80	80	0.021/0.007	Ball fault	4

Table 11. Description of three data sets

Each data set covers four different operating conditions and four different loads (0, 1, 2 and 3 hp). And each class of data sets has 160 data samples which are divided into two equal halves, one for training and the other for testing. The task of data set A is to identify different type of faults, while the experiment over data set B is carried out to further investigate the diagnosis performance of developing faults when the fusion model is trained by incipient faulty samples. And the experiment over data set C is to test the diagnosis performance of incipient faults when the fusion model is trained by the serious faulty samples.

Table 12 gives the results of these three experiments. From Table 12, we can see that the new feature-level fusion model using GEP can get stable, good diagnosis performance. And it is clear that testing performance is higher than training performance in the experiment on data set C. That is to say, when the new feature-level fusion model is trained by the serious faulty samples, it can easily identify incipient faults.

Data set	Training recognition accuracy	Testing recognition accuracy
A	83.75%	81.25%
B	83.75%	72.50%
C	76.25%	81.88%

Table 12. Fault diagnosis performance using feature-fusion model

In order to observe the performance change when the new feature-fusion model uses multiple source information instead of single source information, the new method is used to test bearing fault diagnosis performance with single sensor source. Table 13 gives the performance comparison result between more than one sensor (here using two sensors) and single sensor. From Table 13, we can see multi-sensor testing performance is greatly higher than the single sensor application using the new feature-level fusion model.

Data set	A	B	C
Multi-sensor testing performance increasing	0.56	0.48	0.57

Table 13. Performance comparison between multi-sensor and single sensor

5. Conclusion

This chapter has introduced some new methods for bearing fault diagnosis. These new approaches are using information fusion and intelligent algorithms. Bearing fault diagnosis is still an ongoing research subject over a decade and attracting a huge number of researchers in different areas. But most of those current using techniques mainly deal with single-source data. Many researches have shown that an individual decision system with a single data source can only acquire a limited classification capability which may not be enough for a particular application. So, we study a new way for bearing fault diagnosis using information fusion technology and intelligent algorithm.

Information fusion is a field still under research. Generally, information fusion process may happen in three levels: sensor level, feature level and decision level. Here, we propose a new feature level fusion method and a new decision level fusion method for bearing fault diagnosis. The feature level fusion method is using GEP which is a new intelligent algorithm. And it is a parallel fusion method. The decision level fusion approach is based on a new multiple classifier ensemble method. It analyzes raw vibration signal, and completes the feature extraction by using EMD and fractal feature parameter calculation. From experimental results, we can see that these new fusion model for bearing fault diagnosis task can get good decision performance which is higher than the performance from traditional single sensor application.

Author details

Jiangtao Huang
College of Computer and Information Engineering, Guangxi Teachers Education University
Key Lab of Scientific Computing & Intelligent Information Processing in Universities of Guangxi, China

6. References

[1] Altincay,H. (2005). On naïve Bayesian fusion of dependent classifiers. *Pattern Recognition Letters*, Vol. 26, No. 15, 2463-2473

[2] Bozchalooi, I.S.; Liang, M. (2008). A joint resonance frequency estimation and in-band noise reduction method for enhancing the detectability of bearing fault signals. *Mechanical Systems and Signal Processing*, Vol. 22, 915-933

[3] Case Western Reserve University, Bearing data centre. URL: http://www.eecs.cwru.edu/laborator/bearing

[4] Dara, R.A.; Kamel, M.S.; Wanas, N. (2009). Data dependency in multiple classifier systems. *Pattern Recognition*, Vol. 42, No. 7, 1260-1273

[5] Denoeux, T. (1995). A k-nearest neighbor classification rule based on Dempster-Shafer theory. *IEEE Transactions on Systems, Man and Cybernetics Part B*, Vol. 25, No. 5, 804-813

[6] Ericsson, S.; Grip, N.; Johansson, E.; Persson, L.E.; Sjömberg, J.O. (2005). Towards automatic detection of local bearing defects in rotating machines. *Mechanical Systems and Signal Processing*, Vol. 19, No.3, 509-535

[7] Fan, Y.S.; Zheng, G.T. (2007). Research of high-resolution vibration signal detection technique and application to mechanical fault diagnosis. *Mechanical Systems and Signal Processing*, Vol. 21, No. 2, 678-687

[8] Ferreira, C. (2001). Gene expression programming: a new adaptive algorithm for solving problems. *Complex Systems*, Vol. 13, 87-129

[9] Gu, S.C.; Tan, Y.; He, X.G. (2009). Orthogonal quadratic discriminant functions for face recognition. *Lecture Notes in Computer Science*, Vol. 5553, 466-475

[10] Huang, N.E.; Shen, Z.; Long, S.R.; Wu, M.L.; Shih, H.H.; Zheng, Q.; Yen, N.C.; Tung, C.C.; Liu, H.H. (1998). The empirical mode decomposition and the hilbert spectrum for nonlinear and non-stationary time series anlysis. *Proceedings of the Royal Society of London Series*, Vol. 454, 903-995

[11] Huang, J.T.; Wang, M.H.; Xu, K.K. (2010). Rolling bearing fault diagnosis fusion model based on gene expression programming. *Journal of Information & Computational Science*, Vol. 7, No. 12, 2437-2442

[12] Huang, J.T.; Cao, X.W.; Li, W.J. (2010). Fault diagnosis of rotating machinery based on empirical mode decomposition and fractal feature parameter classification. *Key Engineering Materials*, Vols. 439-440, 658-663

[13] Huang, J.T.; Wang M.H. (2010). Multiple classifiers combination model for fault diagnosis using within-class decision support. *International Conference of Information Science and Management Engineering*, 226–229

[14] Huang, J.T.; Wang, M.H.; Li, W.J. (2011). A new decision-level fusion method for fault diagnosis. *International Journal of Digital Content Technology and its Applications*, Vol. 5, No. 3, 136-142

[15] Junsheng, C.; Dejie, Y.; Yu, Y. (2007). The application of energy operator demodulation approach based on EMD in machinery fault diagnosis. *Mechanical Systems and Signal Processing*, Vol. 21, No. 2, 668-677

[16] Kittler, J.; Hojjatoleslami, A.; Windeatt, T. (1997). Strategies for combining classifiers employing shared and distinct pattern representations. *Pattern Recognition Letters*, Vol. 18, No. 11-13, 1373-1377

[17] Kittler, J.; Hatef, M.; Duin, R.P.W.; Matas, J. (1998). On combining classifiers. *IEEE Transactions on Pattern Analysis and Machine Intelligence*, Vol. 20, No. 3, 226-239

[18] Kuncheva, L.I. (2007). *Combining Pattern Classifiers: Methods and Algorithms*. Wiley-Interscience publication, New Jersey, USA.

[19] Lei, Y.; He, Z.; Zi, Y.; Hu, Q. (2007). Fault diagnosis of rotating machinery based on multiple ANFIS combination with GAs. *Mechanical Systems and Signal Processing*, Vol. 21, No. 5, 2280-2294

[20] Liaw, S.S.; Chiu, F.Y. (2009). Fractal dimensions of time sequences. *Physica A*, Vol. 388, 3100-3106

[21] Mandelbrot, B.B. (1977). *Fractals, Form, Change and Dimension*, Freeman, San Francisco.

[22] Niu, G.; Han, T.; Yang, B.S.; Tan, A.C.C. (2007). Multi-agent decision fusion for motor fault diagnosis. *Mechanical Systems and Signal Processing*, Vol. 21, No. 3, 1285-1299

[23] Ocak, H.; Loparo, K.A.; Discenzo, F.M. (2007). Online tracking of bearing wear using wavelet packet decomposition and probabilistic modeling: a method for bearing prognostics. *Journal of Sound and Vibration*, Vol. 302, No. 4-5, 951-961

[24] Purushotham, V.; Narayanan, S.; Prasad S.A.N. (2005). Multi-fault diagnosis of rolling bearing elements using wavelet analysis and hidden Markov model based fault recognition. *NDT&E International*, Vol. 38, No. 8, 654-664

[25] Rafiee, J.; Rafiee, M.A.; Tse, P.W. (2010). Application of mother wavelet functions for automatic gear and bearing fault diagnosis. *Expert Systems with Applications*, Vol. 37, No. 6, 4568-4579

[26] Rasheed, S.; Stashuk, D.; Kamel, M. (2008). Fusion of multiple classifiers for motor unit potential sorting. *Biomedical Signal Processing and Control*, Vol. 3, No. 3, 229-243

[27] Samanta, B.; Al-Balushi, K.R. (2003). Artificial neural network based fault diagnostics of rolling element bearings using time-domain features. *Mechanical Systems and Signal Processing*, Vol. 17, 317-328

[28] Saravanan, N.; Cholairajan, S.; Ramachandran, K.I. (2009). Vibration-based fault diagnosis of spur bevel gear box using fuzzy technique. *Expert Systems with Applications*, Vol. 36, No. 2, 3119-3135

[29] Tse, P.W.; Zhang, J.Y.; Wang, X.J. (2006). Blind source separation and blind equalization algorithms for mechanical signal separation and identification. *Journal of Vibration and Control*, Vol. 12, No. 4, 395-423

[30] Xu, L.; Krzyzak, A.; Suen, C.Y. (1992). Methods of combining multiple classifiers and their applications to handwriting recognition. *IEEE Transactions on System, Man and Cybernetics*, Vol. 22, No. 3, 418-435

[31] Xu, Z.B.; Xuan, J.P.; Shi, T.L.; Wu, B.; Hu, Y.M. (2009). A novel fault diagnosis method of bearing based on improved fuzzy ARTMAP and modified distance discriminant technique. *Expert Systems with Applications*, Vol. 36, 11801-11807

[32] Yang, J.; Yang, J.Y.; Zhang, D. (2003). Feature fusion: parallel strategy vs. serial strategy. *Pattern Recognition*, Vol. 36, No. 6, 1369-1381

[33] Yang, J.Y.; Zhang, Y.Y.; Zhu, Y.S. (2007). Intelligent fault diagnosis of rolling element bearing based on SVMs and fractal dimension. *Mechanical Systems and Signal Processing*, Vol. 21, No. 5, 2012-2024

[34] Widodo, A.; Yang, B.S.; Han, T. (2007). Combination of independent component analysis and support vector machines for intelligent faults diagnosis of induction motors. *Expert Systems with Applications*, Vol. 32, No. 2, 299-312

[35] Wu, F.; Qu, L. (2009). Diagnosis of subharmonic faults of large rotating machinery based on EMD. *Mechanical Systems and Signal Processing*, Vol. 23, No. 2, 467-475

[36] Wu, J.D.; Chan, J.J. (2009). Faulted gear identification of a rotating machinery based on wavelet transform and artificial neural network. *Expert Systems with Applications*, Vol. 36, No. 5, 8862-8875

Section 3

Magnetic Bearings

Chapter 6

Feasibility Study of a Passive Magnetic Bearing Using the Ring Shaped Permanent Magnets

Teruo Azukuzawa and Shigehiro Yamamoto

Additional information is available at the end of the chapter

http://dx.doi.org/10.5772/51347

1. Introduction

Magnetic bearings can suspend rotating bodies without any mechanical contact. They have advantages such as being free of dust, noise, vibration and maintenance. Some magnetic bearings are already in commercial use in specific apparatuses such as high vacuum pumps or contamination free applications [1]. However, the high cost of the control apparatus for five degrees of freedom of the rotor prevents their wide application at present. It is thus necessary to develop a low-cost magnetic bearing system.

The authors have previously reported on the characteristics of the magnetic force acting between a couple of permanent magnets [2]. A magnetic top, consisting of a couple of ring-shaped permanent magnets, can be levitated without any control while maintaining rotation by itself. This fact suggests that the magnetic top may be a potential candidate for a passive magnetic suspension system. Several efforts have been made to explain the levitation mechanism of the magnetic top. San Miguel proposed noble analytical method with complex formulas showing that a magnetic top can maintain levitation if it rotates with slight precession [3].

In this chapter, an intuitive and easy analytical method based on the equivalent coil currents model for a ring-shaped permanent magnet is proposed.

A quasi-three-dimensional analysis, in which the three-dimensional shapes and layout of the ring-shaped permanent magnets are considered to estimate the magnetic forces acting on the levitating permanent magnet, is proposed. The principle of levitation of the magnetic top and the dimensions of the permanent magnets to realise levitation are discussed using the two-dimensional equations of motion for the magnetic top.

Furthermore, simulations based on the three-dimensional equations of motion are performed to investigate the dynamic behaviour of the magnetic top. The simulated results well predict the dynamic behaviour observed in the experiments. The simulations based on the three-dimensional analysis are used to investigate the effects of the key parameters on the levitating characteristics, such as the sizes of both the ground and rotating permanent magnets, mass of the levitating top, tilt angle of the levitating top, rotation speed and initial position related to the restoring centre.

The ability and feasibility of the magnetic top as a magnetic bearing are also discussed.

2. Analytical methods

The magnetic top is composed of a couple of ring-shaped permanent magnets magnetised in the axial direction, as shown in Figure 1. The magnetic top, equipped with a smaller ring-shaped permanent magnet (a rotor magnet), can be levitated in the magnetic field generated by the larger ring-shaped permanent magnet (a stator magnet) situated at its base, if it can maintain its rotation within a certain speed range. The levitating height is determined by the shapes and magneto-motive forces of the permanent magnets. The authors propose two types of analytical methods: (1) a quasi-three-dimensional analysis to investigate the principle of levitation and the design parameters of the permanent magnets and (2) a three-dimensional dynamic analysis to simulate the behaviour of the levitating magnetic top. The ring-shaped permanent magnet is approximated to the equivalent coil currents model in both the analytical methods.

2.1. The equivalent coil currents approximation

In the equivalent coil currents approximation, a ring-shaped permanent magnet, magnetised in the axial direction, is assumed to exist by the set of circular coil currents located at the outer and inner side surfaces of the ring-shaped permanent magnet [4]. The directions of currents in the outer and inner equivalent coils are inversed with each other, describing the axial magnetization of the permanent magnet, as shown in Figure 2. The magnitude of these equivalent side currents is determined so as to coincide with the measured magnetic field density at the pole surface of the permanent magnet in relation to the number of the assumed equivalent coils.

Figure 3 shows the analytical model based on the equivalent side currents model. The outer and inner diameters and the height of the ring-shaped permanent magnets are represented as d_{so}, d_{si} and h_s for the stator magnet and d_{ro}, d_{ri} and h_r for the rotor magnet, respectively. The angle θ is the tilt angle of the rotor magnet. The origin is set at the centre of the stator magnet. The stator magnet is located in the horizontal x–y plane and z-axis is set as the vertical direction along the axis of the stator magnet. The numbers of the equivalent side currents in both the rotor and stator magnets, indicated as one and three in Figure 3, are decided considering both the accuracy of the calculated results and the required time for computation.

2.2. Magnetic force acting on the rotor magnet

Magnetic force acting on a magnetic top can be estimated by the interaction between the magnetic field generated by the stator magnet and the equivalent coil currents of the rotor magnet. In an analysis based on the equivalent side currents approximation, magnetic forces acting on the magnetic top can be estimated by integrating magnetic forces acting between equivalent coil currents in the rotor and stator magnets.

The magnetic force df [N] acting between two current elements dl_1 [m] and dl_2 [m] and two transporting currents I_1 [A] and I_2 [A] is estimated by the following Biot-Savart's equation:

$$df = \frac{I_1 I_2 dl_1 dl_2}{r^2} \times 10^{-7} \sin\phi \tag{1}$$

where r [m] is the distance between the two current elements and ϕ [rad] is the angle between the directions of the current elements.

Then, the magnetic force f [N] acting between two ring-shaped coil currents is estimated by integrating Equation (1) along the coil sides of the two coil currents l_1 and l_2 as follows:

$$f = \int_{l_1}\int_{l_2} df\, dl_1 dl_2 = I_1 I_2 \int_{l_1}\int_{l_2} \frac{10^{-7}}{r^2} \sin\varphi\, dl_1 dl_2 \tag{2}$$

The magnetic forces acting on the rotor magnet are estimated by integrating Equation (2) for the equivalent side currents. The x, y and z components of the magnetic forces acting on the rotor magnet F_x, F_y and F_z are estimated based on Equation (2).

Figure 1. Experimental magnetic top.

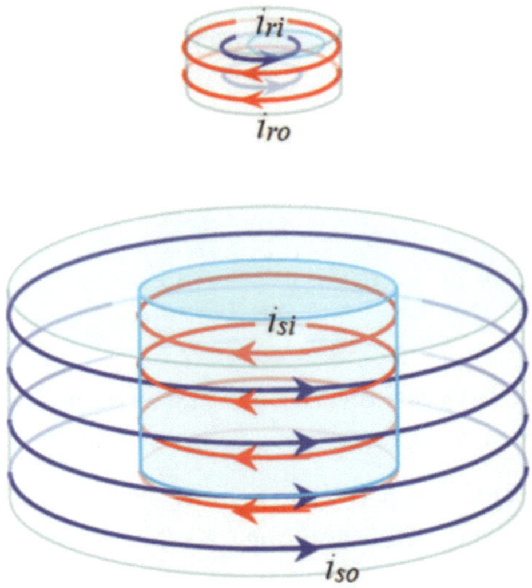

Figure 2. Equivalent side currents.

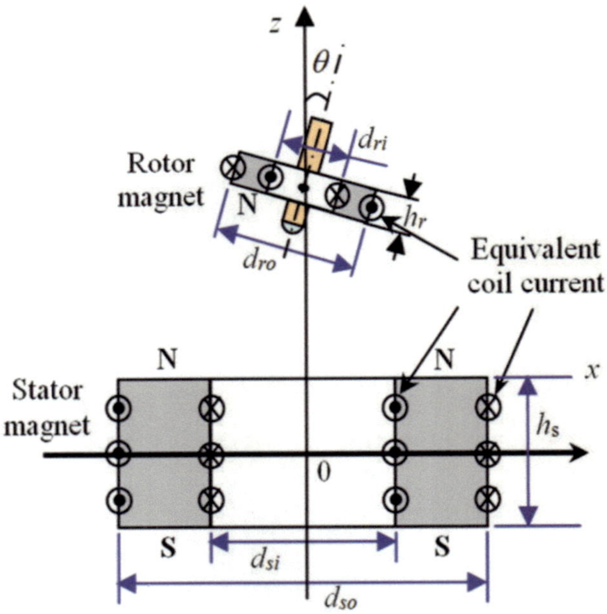

Figure 3. Analytical model.

2.3. Quasi-three dimensional analysis

Because an ideal magnetic top is considered to levitate and rotate around the z-axis, basic information can be obtained by a simple discussion on the two-dimensional motion of the magnetic top in the vertical plane including the z-axis. Hence, the authors propose the quasi-three-dimensional analysis in which the magnetic force acting on the rotor magnet is estimated using Equations (1) and (2), considering the circular shapes and layout of the equivalent coil currents. The behaviour of the levitating magnetic top in z–x plane is estimated by the following two-dimensional equations of motion for the rotor magnet:

$$F_x = m\frac{d^2x}{dt^2} \tag{3}$$

$$F_z = m\left(\frac{d^2z}{dt^2} + g\right) \tag{4}$$

where F_x and F_z [N] are the magnetic forces acting on the rotor magnet in x and z directions, respectively, m [kg] is the mass of the magnetic top, g [m/s^2] is the acceleration due to gravity and (x, z) are the coordinates of the centre of the rotor magnet. Here, Equation (4) indicates that the vertical acceleration is derived from the difference between the vertical component of the magnetic force due to the stator magnet and the gravity force acting on the rotor magnet.

The quasi-three-dimensional analysis is used to investigate the principle of levitation of the magnetic top and determine with a short computing time the parameters of the magnetic top such as the sizes of the stator and rotor magnets and the levitation height.

2.4. Three-dimensional dynamic analysis

Because the quasi-three-dimensional analysis provides the design parameters of a magnetic top, behaviour of the magnetic top is investigated by a simulation based on three-dimensional dynamic analysis considering rotation of the magnetic top. Behaviour of the magnetic top can also be estimated by the equations of motion on the angular moment of the rotor magnet, considering three-dimensional layout of the stator and rotor magnets, the tilt angle of the rotor magnet and the mechanical inertia of the rotor magnet.

When the magnetic top is rotating, the angular momentum force I_{Top} will act around the axis of the rotating magnetic top. The momentum force I_{Top} can be expressed as Equation (5), where m is the mass of the levitating magnetic top, r_{ro} and r_{ri} are the outer and inner radius of the ring-shaped rotor magnet. Angular momentum vector around the axis of the rotating magnetic top L_n at a time t_n can be expressed as Equation (6), where ω is the angular velocity of the magnetic top. The incremental angular momentum $d\vec{L}$ in an infinitesimal time dt is expressed as Equation (7), where $\vec{N}$ is the moment caused by magnetic force acting on the rotor magnet. The angular momentum vector $\vec{L}_{n+1}$ at time $t_{n+1}=t_n + dt$ is expressed as Equation (8):

$$I_{Top} = m\left(r_{ro}^{\ 2} + r_{ri}^{\ 2}\right)/2 \tag{5}$$

$$L_n = I_{Top} \times \omega \tag{6}$$

$$d\vec{L} = dt \times \vec{N} \tag{7}$$

$$\vec{L}_{n+1} = \vec{L}_n + d\vec{L} \tag{8}$$

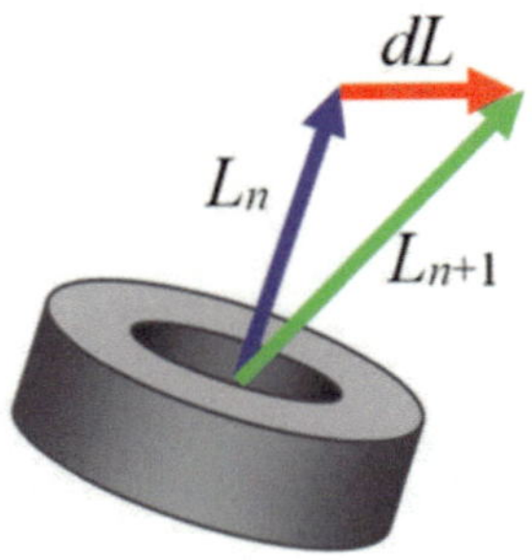

Figure 4. Angular momentum of a top

The moment $\vec{N}$ in Equation (7) corresponds to torque and can be estimated by the magnetic force acting on the rotor magnet with a calculation based on the equivalent coil currents model. The motion of the magnetic top can be simulated using the following equations by the 4th order Runge-Kutta method:

$$t_{n+1} = t_n + h \tag{9}$$

$$k_1 = h\, f(t_n,\, v_n) \tag{10}$$

$$k_2 = h\, f(t_n+h/2,\, v_n+k_1/2) \tag{11}$$

$$k_3 = h\, f(t_n+h/2,\, v_n+k_2/2) \tag{12}$$

$$k_4 = h\, f(t_n+h,\, v_n+k_3) \tag{13}$$

$$v_{n+1} = v_n + (k_1 + 2\,k_2 + 2\,k_3 + k_4)/6 \tag{14}$$

where h is the incremental time. In this analysis, the aerodynamic damping effects are neglected for easy calculation.

3. Levitating characteristics of a magnetic top based on quasi-three-dimensional analysis

3.1. Principle of levitation of a magnetic top

To investigate levitation characteristics intuitively, the authors have proposed a so-called 'magnetic force map' that shows the magnetic force acting on the rotor magnet at each mesh point in the magnetic field generated by the stator magnet. Magnetic forces at the mesh points above the stator magnet are shown in the vector diagram. Because the vertical component of the magnetic force is deducted by the weight of the levitating top, we can observe the net force acting on the top at a glance.

Table 1 shows the parameters of the analytical model used in this chapter. These parameters are for the experimental model introduced in Figure 1. The magnitude of current in each equivalent side current is determined to be equal to the magnetic field density at the surface of the permanent magnets and the measured values for the ferrite permanent magnets used in the experiments. Considering the thickness of the permanent magnets, the number of the equivalent current coils is set to be 2 for the rotor magnet and 24 for the stator magnet in the simulation. Each circular coil current is simulated as a set of 72 linear current elements. These parameters are determined considering the accuracy of calculated results and the required time for computation.

	Rotor magnet	Stator magnet
Outer diameter d_o [mm]	30	134
Inner diameter d_i [mm]	12	75
Thickness h [mm]	5	60
Magnitude of equivalent current I_{eq} [A/mm]	286	286
Mass m [g]	20.37	-
Tilt angle θ [deg]	-	1
No. of equivalent coils	2	24
No. of current elements in an equivalent coil	72	72

Table 1. Parameters used in simulation

Figure 5 shows the magnetic force map calculated for the parameters given in Table 1. The figure shows the distribution of the magnetic force acting on the rotor magnet at each mesh point in the vertical plane including the z–x plane. Although the magnetic force map displays the force distribution in a two-dimensional plane, the magnetic forces are calculated considering three-dimensional shapes and layout of the equivalent side currents.

Figure 5(a) shows the magnetic force map in case the tilt angle of the rotor magnet is zero, that is, the rotor magnet is laid out horizontally in the area above the stator magnet. This figure shows that the force distribution is not uniform in the space above the stator magnet. There are two singular points along the z-axis: points A (0, 99.5) and B (0, 91.5) (Figure 5(a)). At point A, the magnetic forces acting on the rotor magnet are stable in the vertical direction but unstable in the horizontal direction. On the contrary, at point B, the magnetic forces acting on the rotor magnet are unstable in the vertical direction but stable in the horizontal direction. These results show that the magnetic top cannot levitate when its axis is parallel to the vertical axis; this result accords with the Earnshaw's theorem.

Figure 5(b) shows the magnetic force map when the tilt angle of the rotor magnet θ is set to 1° in $x < 0$ to −1° in $x > 0$. This figure shows that there is a point where the magnetic forces acting on the rotor magnet are stable in the both horizontal and vertical directions, as shown by the point C (0, 99.5) in Figure 5(b). In other words, the magnetic forces will guide the rotor magnet to the equilibrium point C, named as the 'restoring centre' in this chapter.

The quasi-three-dimensional analysis shows that there is no restoring centre when the tilt angle of the rotor magnet is 0, but a slight tilt angle such as 1° brings the restoring centre into existence. These results suggest that a magnetic top equipped with a ring-shaped permanent magnet can levitate in the space above a stator ring-shaped permanent magnet if it rotates with a slight precession.

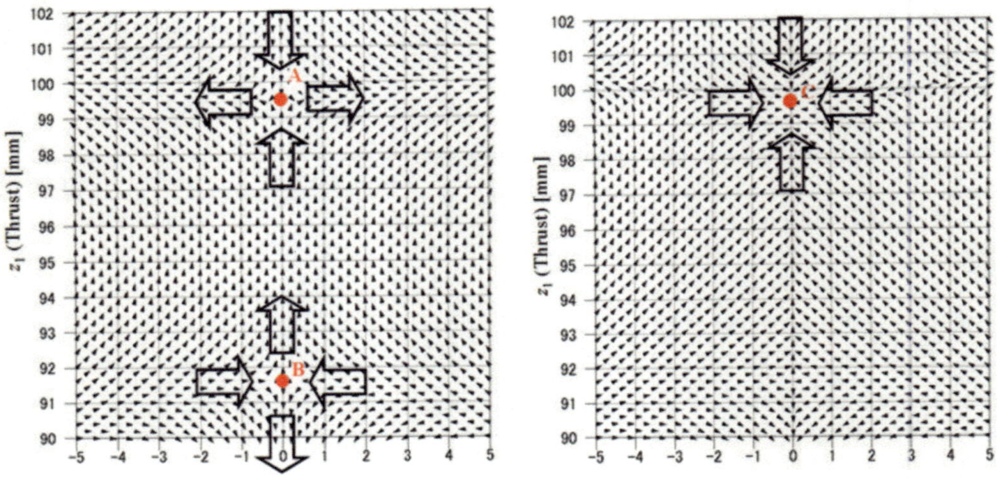

(a) $\theta = 0°$, rotor magnet is horizontal. (b) $\theta = 1°$ with respect to z-axis.

Figure 5. Magnetic force map for different tilt angles θ of a levitating magnetic top.

3.2. Simulation to investigate the behaviour a magnetic top

To confirm the validity and effectiveness of quasi-three-dimensional analysis using the magnetic force map, dynamic behaviour of the rotor magnet is investigated by computer simulation based on the equations of motion introduced in the previous section. To make intuitive discussions, a dynamic simulation using two-dimensional equations of motion, Equations (3) and (4), is performed. In this simulation, the tilt angle of the rotor magnet is set to $\theta = 1°$ in the area $x < 0$ and to $-1°$ in the area $x > 0$.

Figure 6 shows the simulated behaviour of the centre of the rotor magnet for 10 s starting from the point (1, 98.5), which is 1 mm apart in both x and z directions from the restoring centre (0. 99.5). The simulated time trajectory of the centre of the rotor magnet (Figure 6(a)) shows that the rotor magnet levitates in the area of ±1 mm in both vertical and horizontal directions from the restoring centre. The bottom left point of this rectangular space is the initial position of the rotor magnet. These results tell us that the magnetic top is swaying around the restoring centre and the range of swaying motion is determined by the initial position of the magnetic top with regard to the restoring centre. Figures 6 (b) and (c) show the time dependencies of radial and vertical motions of the centre of the rotor magnet. From these figures, we find that the frequencies of radial and vertical motions are 1.45 Hz and 1.13 Hz, respectively.

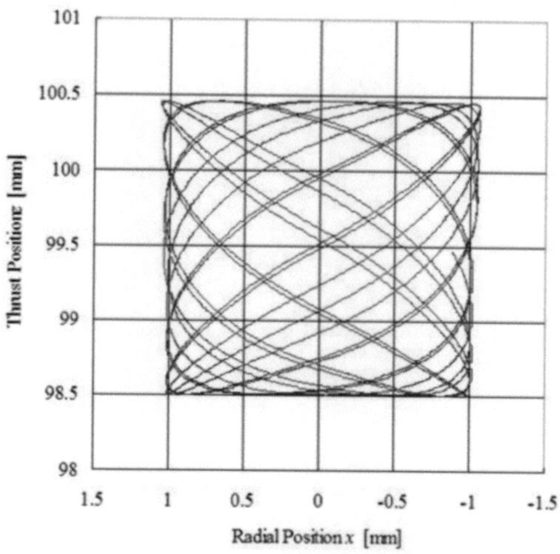

(a) Trajectory of the centre of the rotor magnet.

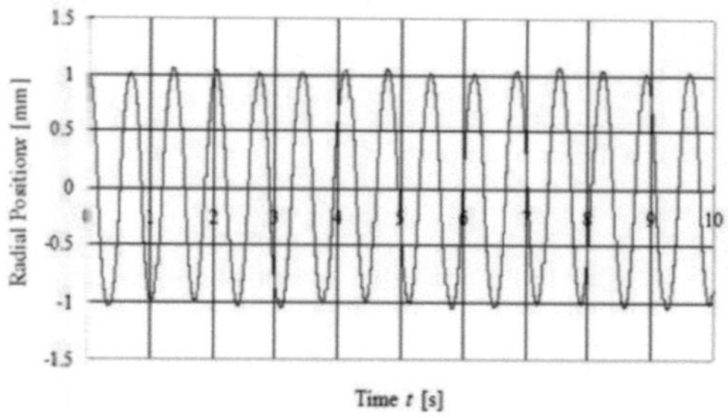

(b) Time dependency of radial motion.

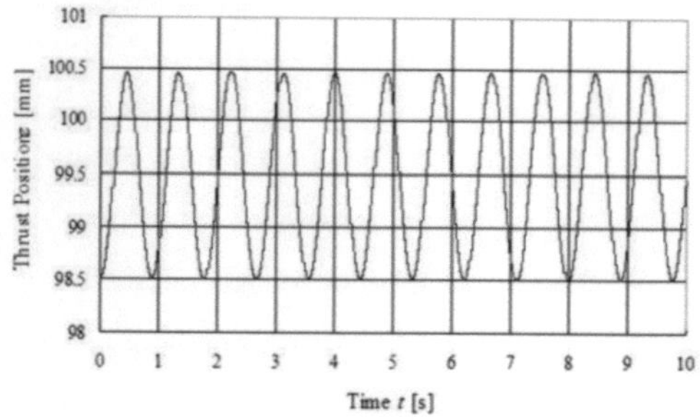

(c) Time dependency of vertical motion

Figure 5. Simulated behaviour of the centre of the rotor magnet based on two-dimensional analysis.

3.3. Validity of the quasi-three-dimensional analysis

To verify the validity of the above analytical results, experiments are performed using the test model. The dimensions of the rotor and stator magnets used in the test model are listed in Table 1. The weight of the top is adjusted to 20.37 g using a dummy weight. Behaviour of the levitating magnetic top is recorded using a video camera from the y direction. The levitation height of the centre of the rotor magnet is about 100 mm above the centre of the stator magnet. The digital image information is obtained using motion capture software 'Pv Studio 2D demo' and the software 'Graph Scan 1.8' are used to obtain Figure 7. The frame size and frame interval of the obtained video data are 640 × 480 pixels and 30 flames per second. However, finally obtained frame interval using the above software is 4 frames per second.

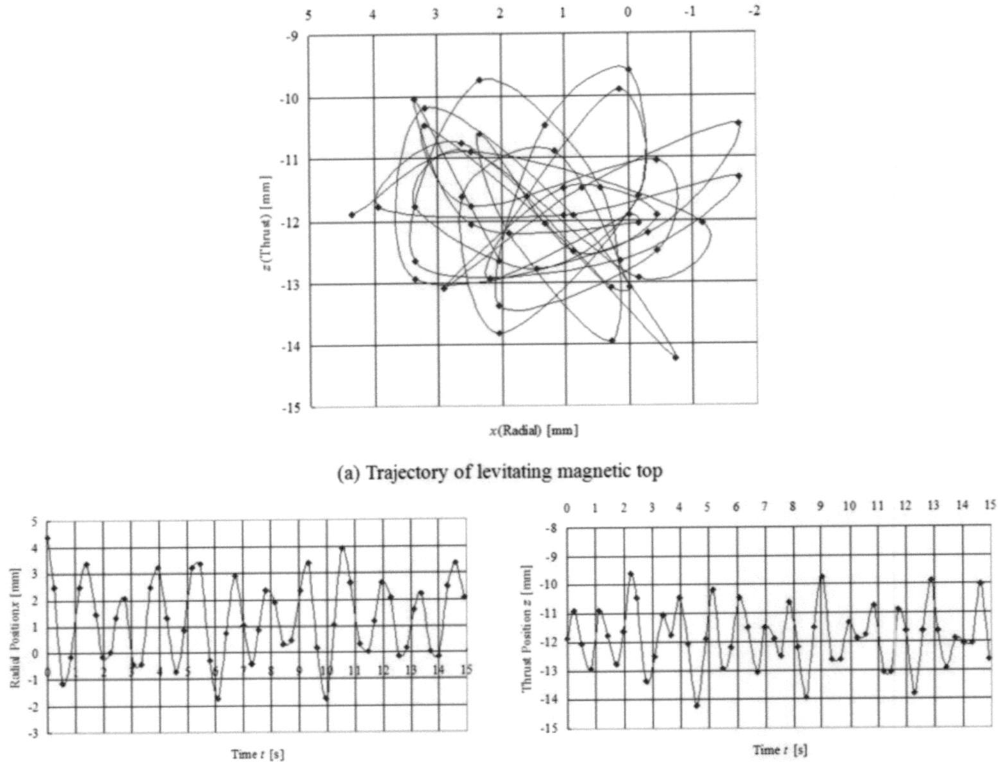

(a) Trajectory of levitating magnetic top

(b) Time dependence of radial position of the top

(c) Time dependence of thrust position of the top

Figure 7. Measured behaviour of the magnetic top in the test model.

Experiment is performed according to the following steps : (1) place a non-magnetic plate on the pole surface of a stator magnet, (2) rotate a magnetic top on the plate at the centre of a stator magnet, (3) lift the plate with rotating top slowly until a magnetic top is pulled into the restoring centre. There are some hurdles to clear these steps. A magnetic top should be rotate at the exact centre of the stator magnet in a certain rotating speed range to clear step (2). Lift force should be less than vertical magnetic force acting on a top from the stator magnet to clear step (3). A magnetic top is rotated by fingers and the plate is lifted by hand in our experiment. Then, it is difficult to obtain experimental data of the same conditions.

Figure 7 shows the measured trajectory of the centre of the rotor magnet for 15 s. In Figure 7, the origin of x and z coordinates is the centre of the picture captured by the camera. In Figure 7(a), dots indicate the positions of the rotor magnet centre measured every 0.25 s, *i.e.* 4 frames per second, and a smoothing line connects these dots in sequential order. The smoothing line in Figure 7(a) does not show the swaying motion correctly; however, we can observe that the rotor magnet levitates and sways in the range of ±3 mm in radial direction and ±2.3 mm in vertical direction. In the experiment, it is difficult to start rotation of the magnetic top at the designated initial point. Figure 7(a) suggests that the initial positions of the rotor magnet in this experiment were 3 mm and 2.3 mm apart from the restoring centres in x and z directions, respectively. Figures 7 (b) and (c) demonstrate the time dependence of the radial and vertical motions in 15 s. These figures show that the frequencies of swaying motion are about 0.75 Hz in radial direction and about 1.05 Hz in vertical direction. These test results are compared to the calculated ones in Table 2.

	Measured	Calculated
Levitation height [mm]	100	99.5
Frequency of radial swaying [Hz]	0.75	1.45
Frequency of vertical swaying [Hz]	1.05	1.13

Table 2. Comparison between analysis and experimental data

In spite of low accuracy of the measured data and difficulties in reenacting experiments in the same condition, the analysed levitation height and the frequency of vertical swaying are well in accordance with the experimental values. However, the analysed frequency of radial swaying is about twice the experimental value. This difference seem to be derived from assumptions in the two-dimensional analysis such as the constant tilt angle of the rotor magnet. Simulated results for various tilt angles showed that the magnitude of the tilt angle significantly affects the radial motion, but does not affect the vertical motion of the rotor magnet. Furthermore, the analysis is based on two-dimensional equations of motion, and three-dimensional behaviour of the magnetic top in the experiment is measured as two-dimensional video information.

These results show that the fundamental parameters of a magnetic top, such as levitation height and dimensions of the permanent magnets, can be determined well using the quasi-three-dimensional analysis.

3.4. Levitating area and parameters of the magnets

Figure 8 shows the magnetic force map for the test model shown in Table 1. The tilting angle of the rotor magnet is set as ±1°. This figure shows that a tilting magnetic top, located within the red dotted lines and named as the 'levitating area', will be guided by the magnetic force along the direction of vectors towards the restoring centre A (0, 0, 99.5). Although the levitating area is shown as a two-dimensional area in this figure, the real shape of the levitating area is conic.

The size and shape of the levitating area are closely related to the dimensions of the permanent magnets and the tilting angle of the rotor magnet. Figure 9 shows the relationship between the shape and size of the levitating area and the parameters of the permanent magnets. The effects of precession are considered to set the tilt angle θ to −1° in $x > 0$ and to 1° in $x < 0$. The conical shape of the levitating area is approximated by the rectangular area bounded by the green coloured dotted line in Figure 8 [5].

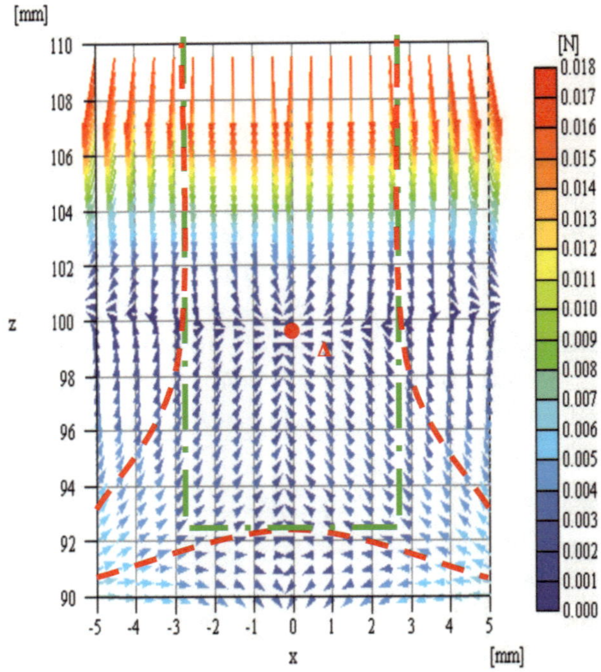

Figure 8. Levitating area of a magnetic top tilted by 1°.

Figure 9(a) shows the levitating areas and the restoring points for the various inner diameters of the rotor magnet d_{ri}. When the inner diameter of the rotor magnet increases, the restoring point becomes higher and the levitating area becomes narrower in the radial direction and wider in the thrust direction. These results indicate that relatively well radial bearing characteristics can be obtained by a rotor magnet with a large inner diameter. On the contrary, relatively well thrust bearing characteristics can be obtained by a rotor magnet with a smaller inner diameter.

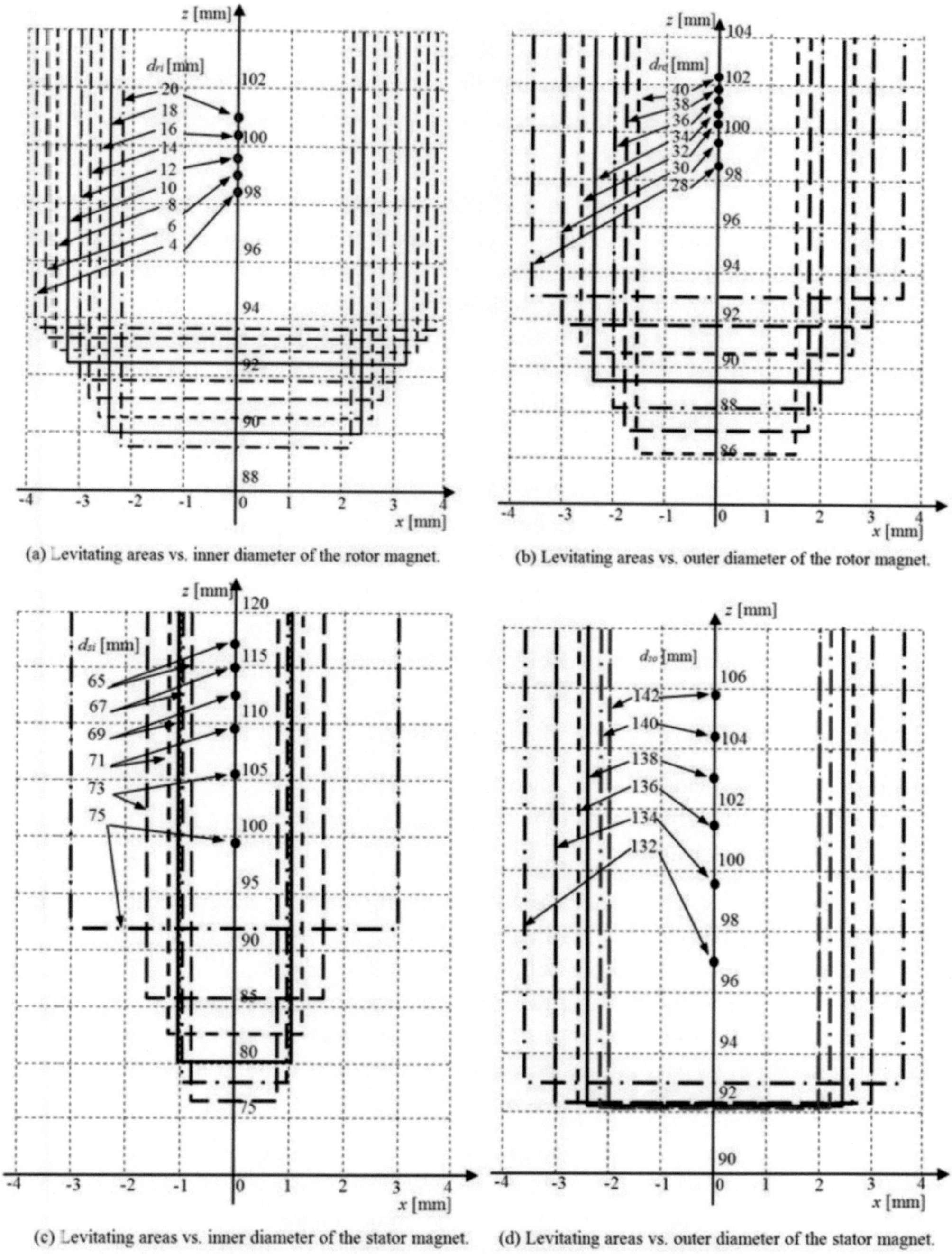

(a) Levitating areas vs. inner diameter of the rotor magnet.

(b) Levitating areas vs. outer diameter of the rotor magnet.

(c) Levitating areas vs. inner diameter of the stator magnet.

(d) Levitating areas vs. outer diameter of the stator magnet.

Figure 9. Relationship between the shape and size of the levitating area and the parameters of the permanent magnets.

Figure 9(b) shows the levitating areas and the restoring points in the case where the outer diameter of the rotor magnet d_{ro} changes. When the outer diameter of the rotor magnet increases, the restoring point becomes higher and the levitating area becomes narrower in the radial direction and wider in the thrust direction. These results state that relatively well radial bearing characteristics can be obtained by a rotor magnet with a large outer diameter. On the contrary, relatively well thrust bearing characteristics can be obtained by a rotor magnet with a smaller outer diameter.

Figure 9(c) shows the levitating areas and the restoring points in case where the inner diameter of the stator magnet d_{si} changes. When the inner diameter of the stator magnet increases, the restoring point becomes lower and the levitating area becomes wider in the radial direction and narrower in the thrust direction. These results show that relatively well radial bearing characteristics can be obtained by a stator magnet with a smaller outer diameter. On the contrary, relatively well thrust bearing characteristics can be obtained by a stator magnet with a larger outer diameter.

Figure 9(d) shows the levitating areas and the restoring points in the case where the outer diameter of the stator magnet d_{so} changes. When the outer diameter of the stator magnet increases, the restoring point becomes higher and the levitating area becomes narrower in the radial direction. The outer diameter of the stator magnet hardly affects the axial height of the levitating area. These results indicate that relatively well radial bearing characteristics can be obtained by a stator magnet with a large outer diameter. The thrust bearing characteristics are not changed by the outer diameter of the stator magnet.

3.5. Levitating area and tilt angle of the magnets

As mentioned in the previous section, the tilting of a rotor magnet is essential in a magnetic top. In this section, relations between the shapes of the levitating area and the tilt angle of the rotor magnet are discussed.

Figure 10 shows the magnetic force map with levitating areas for different tilt angles of the rotor magnet. Figure 10(a) shows the magnetic force map when the tilt angle of the rotor magnet is zero. The magnetic forces acting on the rotor magnet are stable in the vertical direction but unstable in the radial direction at the upper singular point (0, 99.5). On the contrary, the magnetic forces acting on the rotor magnet are unstable in the vertical direction but stable in the radial direction at the lower singular point (0, 91.5). In this case, there is no levitating area because there is no restoring centre.

Figure 10(b) shows the magnetic force map when the tilt angle of the rotor magnet is $\theta = 0.4°$, *i.e.* $\theta = 0.4$ in the area $x < 0$ and $\theta = -0.4$ in the area $x > 0$. Figures 10(c) and (d) show the magnetic force maps when the tilt angle of the rotor magnet is $\theta = 0.8°$ and $\theta = 1.2°$, respectively. Figures 10 and 8, showing the case of $\theta = 1.0°$, illustrate the fact that the levitating area becomes wider when the tilt angle becomes larger up to 1.2°, while the magnitude of restoring force around the restoring centre becomes saturated. We can intuitively observe considering the behaviour of a normal top that a magnetic top with a very large tilting angle will not levitate. Figure 10 also shows that the height of the restoring centre does not change by the tilt angle of the rotor magnet.

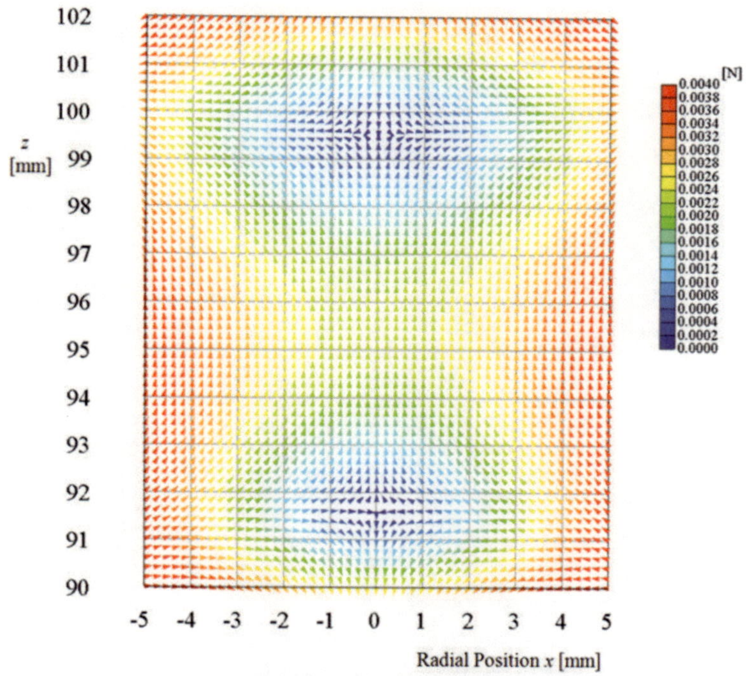

(a) The magnetic force map when the tilt angle is zero.

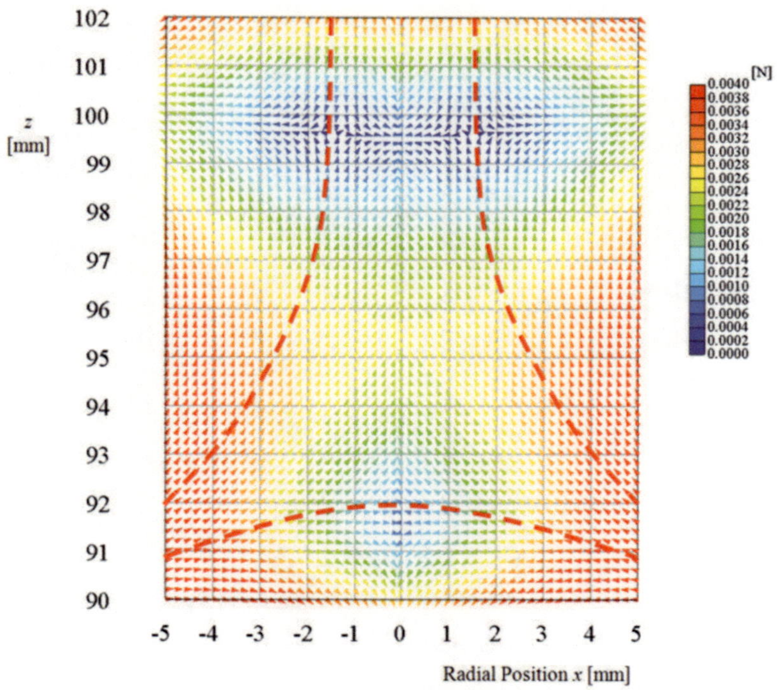

(b) The levitating area when the tilt angle is 0.4°.

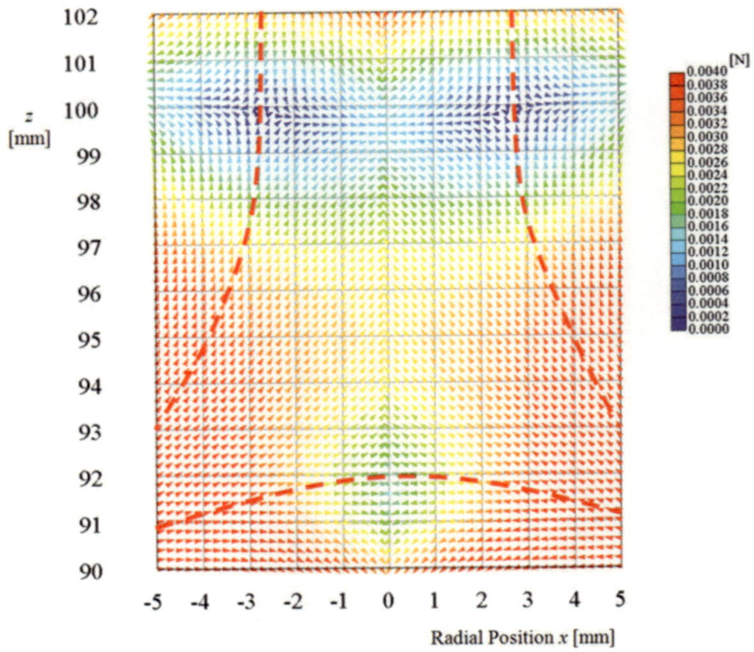

(c) The levitating area when the tilt angle is 0.8°.

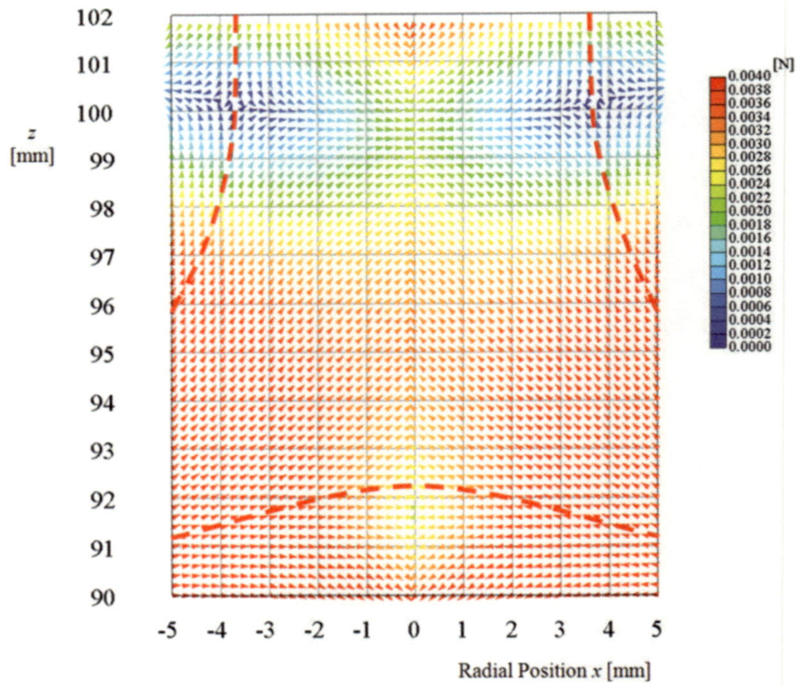

(d) The levitating area when the tilt angle is 1.2°.

Figure 10. Relationship between the levitating area and the tilt angle of the rotor magnet.

4. Study of the dynamic behaviour of a magnetic top by three-dimensional analysis

We can obtain approximate guidelines for the size and shape of the levitating area by quasi-three-dimensional analysis. Although the static analysis gives the 'levitating area', a magnetic top in this area may not always continue to levitate, considering the dynamic motion of the top. Furthermore, the static analysis mentioned in the previous section showed that the magnitude of the restoring force acting on the rotor magnet was small. Because the quasi-three-dimensional static analysis provides an approximate design of the magnetic top, the three-dimensional dynamic analysis should be performed to confirm whether the rotating magnetic top can maintain levitation.

In this section, how the parameters such as rotating speed, mass of the top and initial position with regard to the restoring centre affect the behaviour of the levitating magnetic top is discussed.

4.1. Effects of rotating speed

To realise a successful rotation of a magnetic top, the rotation speed is one of the most important parameters. Simulated results show that the magnetic top (Table 1) can maintain levitating while it rotates in the range of 18–50 rps, *i.e.* 1,080–3,000 rpm, when the initial position is 1 mm apart in both radial and vertical directions from the restoring centre.

Figures 11(a) and (b) show the trajectories of the centre of the magnetic top rotating at 1020 rpm and 3240 rpm, respectively. This characteristic is closely related to the tilt angle of the rotor magnet, that is, the rotor magnet with the shaft rotating at very low speed cannot maintain an adequate tilt angle because of the lack of mechanical inertia and the rotor magnet with the shaft rotating at a very high speed cannot maintain its tilt angle stable because of the increasing centrifugal force.

Figure 12 shows the typical time dependency of the tilt angle of the rotor magnet. The tilt angle in this figure indicates the absolute values, *i.e.* the rotor magnet is tilting in a radial direction around z-axis. As shown in this figure, the tilt angle θ varies within 1.2° while the rotor magnet levitates with precession, as in this case. The maximum value of the tilt angle increases with increase in the rotation speed of the rotor magnet, as shown in Figure 13. In this analytical model, the gravity centre of the magnetic top is located at a little upper point along its shaft from the centre of the rotor magnet; therefore, the tilt angle becomes larger with an increase in the rotating speed. Then, the rotor magnet will be thrown in the radial direction, along the magnetic force vectors shown in Figure 10(a). If we design a magnetic top with the gravity centre located at the centre of the rotor magnet, the rotor magnet will rotate without tilting because of its mechanical inertia; however, such a rotor magnet cannot realise levitation, according to previous discussions.

Figure 14 shows the simulated trajectories of a levitating magnetic top rotating at 1,080 rpm for 60 s after starting from point (1, 0, 98.5), which is 1 mm apart from the restoring centre in both x and z directions. Figures 14(a) and (b) show the trajectories of the head of the 25 mm long shaft of the magnetic top and Figures 14(c) and (d) show the trajectories of the centre of the rotor magnet. Figures 14(a) and (b) show that the shaft head rotates with both smaller radius nutation and larger radius precession. On the other hand, Figures 14(c) and (d) show that the centre of the rotor magnet rotates with precession when the tilt angle varies periodically. Comparing these two figures, it is observed that a magnetic top, rotating at a low speed such as 1,080 rpm, is rotating in a complex motion with nutation mode in addition to precession mode [6].

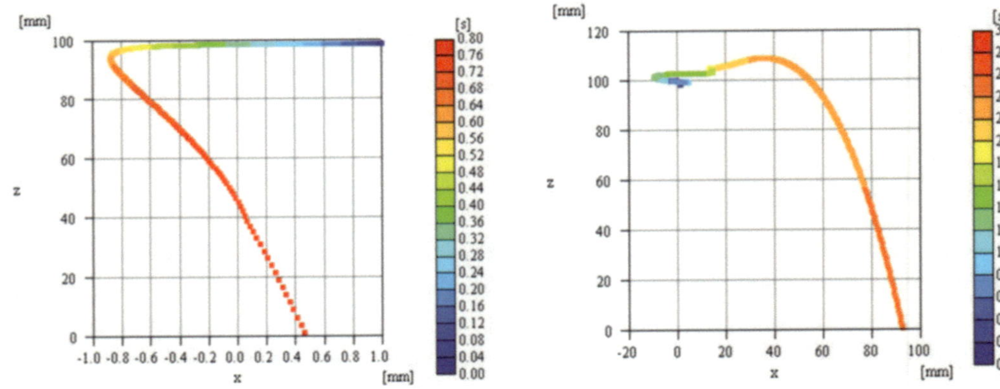

(a) Rotating speed: 1020 rpm (b) Rotating speed: 3240 rpm

Figure 11. Trajectories of the magnetic top for 5 s.

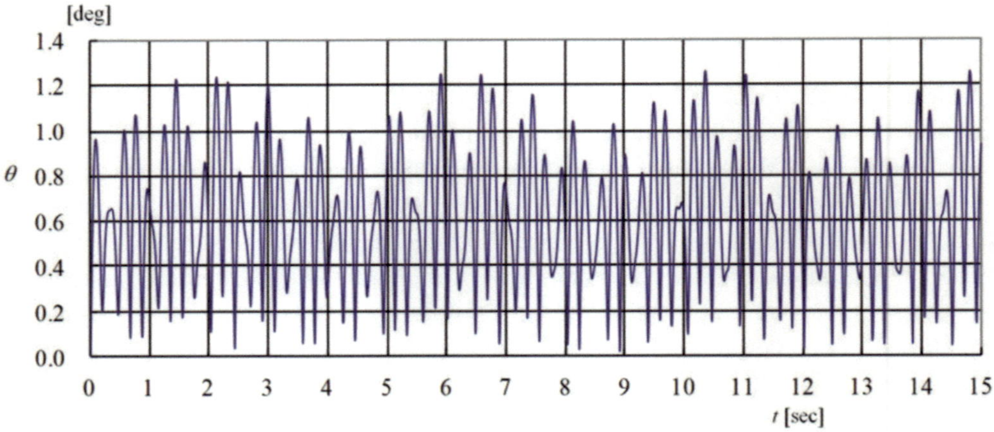

Figure 12. Time dependency of the tilt angle of the rotor magnet.

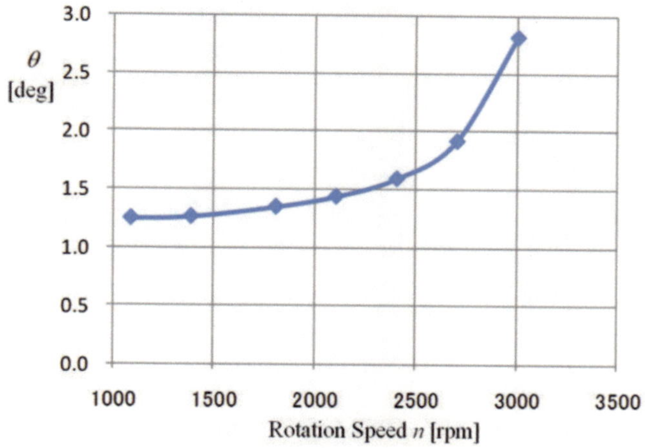

Figure 13. The maximum tilt angle θ vs. rotating speed.

Figure 15 shows the simulated trajectories of a levitating magnetic top rotating at 3,000 rpm for 60 s after starting at point (1, 0, 98.5). Figures 15(a) and (b) show the trajectories of the head of the 25 mm long shaft of the magnetic top and Figures 15(c) and (d) show the trajectories of the centre of the rotor magnet. From these figures, we can observe that both the trajectories of the shaft head and the centre of the rotor magnet are almost the same in shape. However, the shaft head rotates in a little wider range compared to the moving area of the centre of the rotor magnet. This means that a magnetic top rotating at a relatively higher speed, e.g. 3,000 rpm, maintains its levitation with precession mode. In this case, nutation mode is hardly observed.

Although it is difficult to repeat the experiments in the same conditions, these simulated results showed good accordance with the experiments [6].

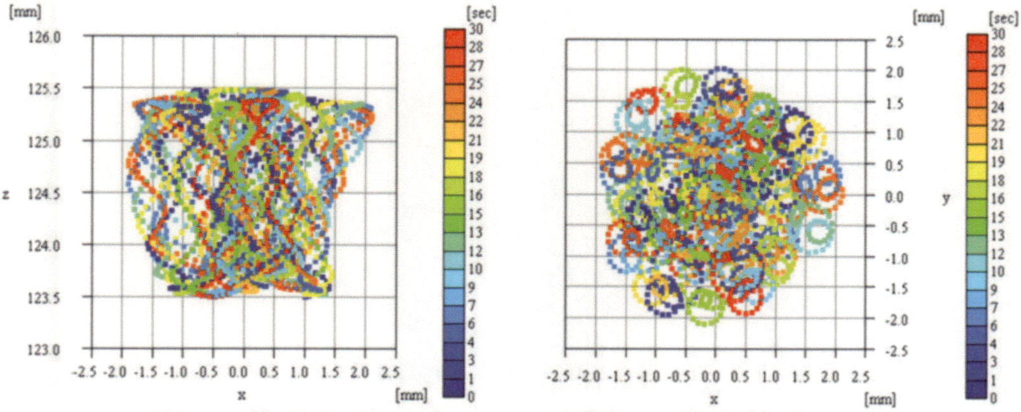

(a) Trajectory of the shaft head in z–x plane.

(b) Trajectory of the shaft head in y–x plane.

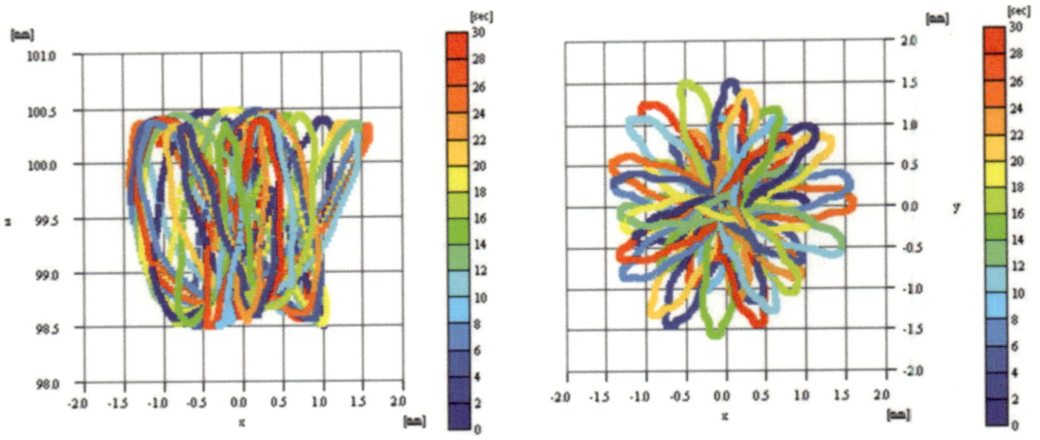

(c) Trajectory of the levitating magnet centre in z–x plane. (d) Trajectory of the levitating magnet centre in y–x plane.

Figure 14. Simulated trajectories of the levitating magnetic top rotating at 1,080 rpm.

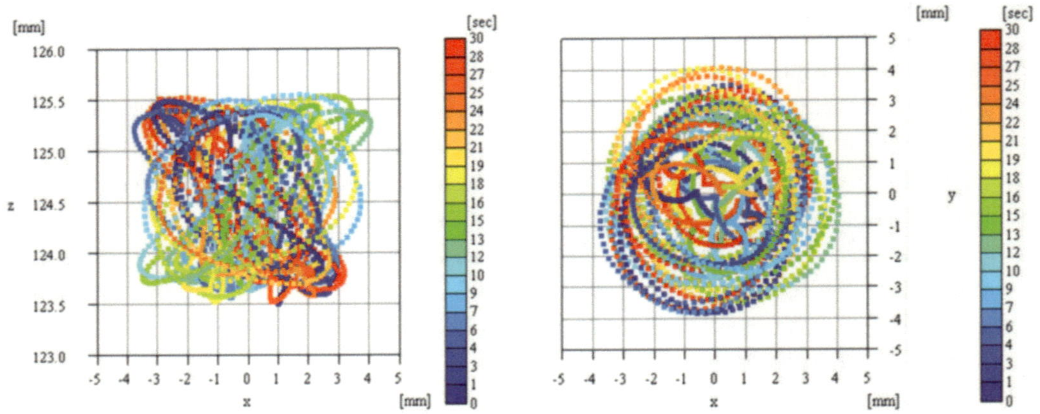

(a) Trajectory of the shat head in z–x plane. (b) Trajectory of the shaft head in y–x plane.

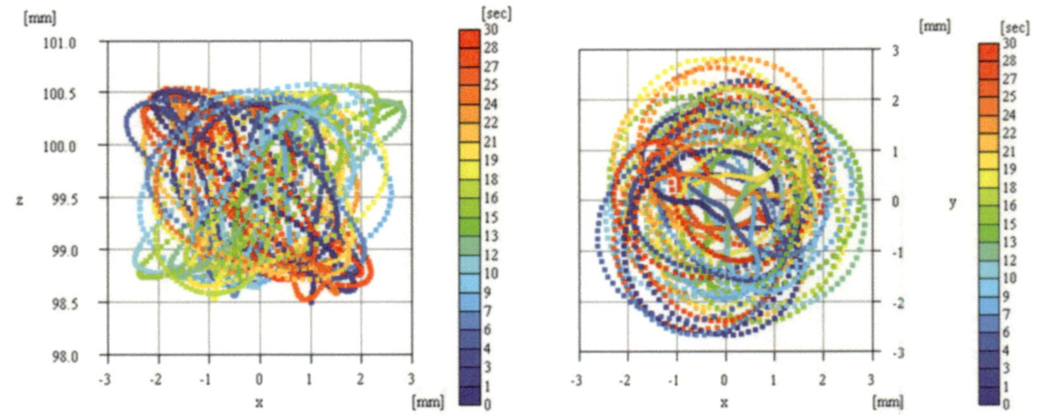

(c) Trajectory of the magnetic centre in z–x plane (d) Trajectory of the magnetic centre in y–x plane

Figure 15. Simulated trajectories of the levitating magnetic top rotating at 3,000 rpm.

4.2. Effects of the mass of a magnetic top and thickness of the stator magnet

In this section, the effects of the mass of a levitating top and the thickness of the stator magnet to the height of the restoring centre are discussed.

Figure 16 shows the relationship between the height of the restoring centre z_r [mm] and the mass of a levitating top m [g] when the thickness of the stator magnet is h = 60, 40 and 20 mm. Calculated results show that the height of the restoring centre z_r decreases with an increase in the mass of the top and a decrease in the thickness of the stator magnet.

Calculations were made for various values of the mass of the top in the analytical model described in Table 1. However, there is no restoring centre or levitation area for a heavier or a lighter top than those shown in Figure 16. According to these results, a thin stator magnet may realise successful levitation for wider mass variations.

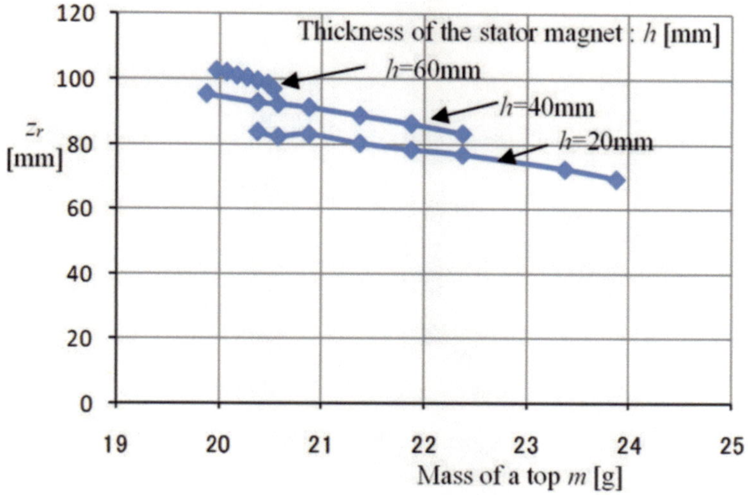

Figure 16. Height of the restoring centre vs. mass of levitating top.

4.3. Effects of the initial position

Some experiments demonstrated that the initial position related to the restoring centre is one of the most important parameters. To realise successful rotation of a magnetic top, the initial position should be at least inside the levitating area defined in the previous section. The magnetic top shows various behaviours according to its initial point with regard to the restoring centre.

Figure 17 shows the simulated trajectory of the centre of the rotor magnet for 60 s starting from (1, 0, 98.5), which is 1 mm apart in both x and z directions from the restoring centre. The mass of the top is 20.37 g and rotation speed is 23 rps, *i.e.* 1380 rpm. These results show that the rotating top is levitated in the area of ±1.6 mm in both x and y directions and of ±1 mm in z direction, from the restoring centre. The maximum tilt angle is 1.26° with the z-axis.

To investigate the effects of the initial point with regard to the restoring centre (0, 0, 99.5), simulations were performed for the case of the typical initial point of (1, 0, 99.5), *i.e.* 1 mm apart in x direction from the restoring centre, and (0, 0, 98.5), *i.e.* 1 mm apart in z direction from the restoring centre. Figures 18(a) and (b) show the simulated trajectories of the centre of the rotor magnet, rotating at 1,380 rpm for 60 s starting from (1, 0, 99.5) and (0, 0, 98.5), respectively.

The magnetic top, starting from the point 1 mm apart in x direction from the restoring centre, levitates in the range of ±1.07 mm in both x and y directions and from +0.12 mm/ to 0.06 mm in z direction around the restoring centre, as shown in Figure 18(a). The maximum tilt angle is 1.018°.

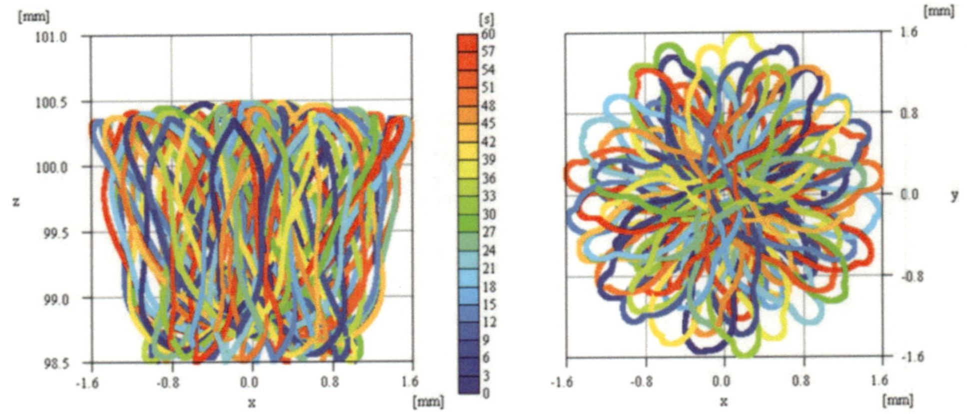

(a) Trajectory in z–x plane (b) Trajectory in y–x plane

Figure 17. Simulated trajectories of a magnetic top in 60 s starting from (1, 0, 98.5), 1380 rpm.

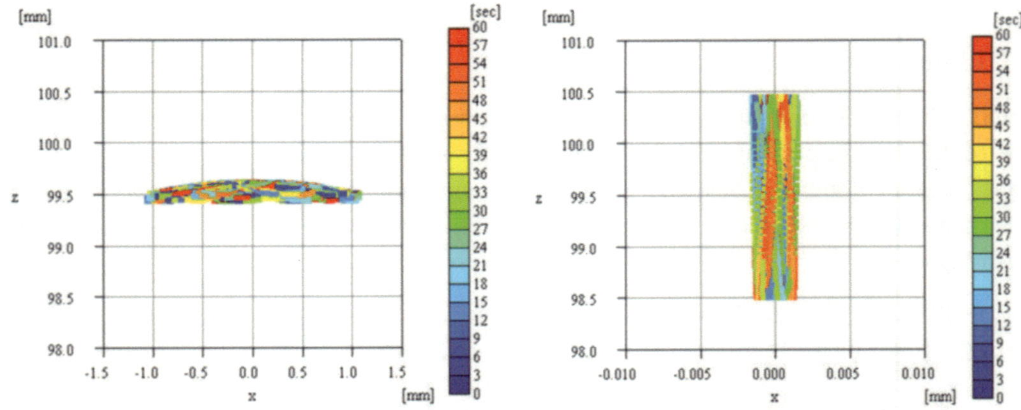

(a) Trajectory of the rotor magnet starting from (1, 0, 99.5) (b) Trajectory of the rotor magnet starting from (0, 0, 98.5)

Figure 18. Trajectories of the centre of the rotor magnet in z–x plane.

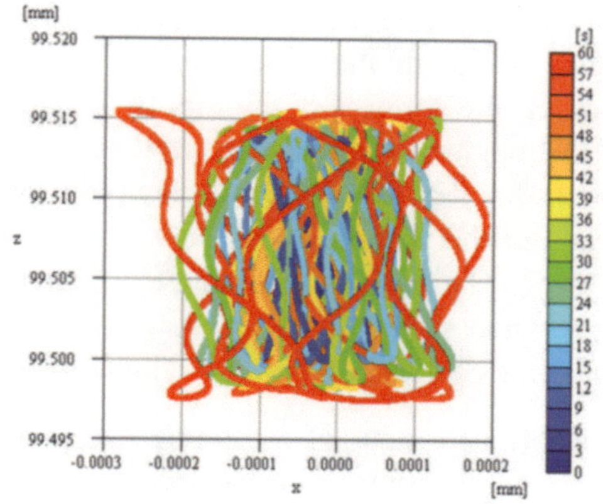

(a) Trajectory in z–x plane

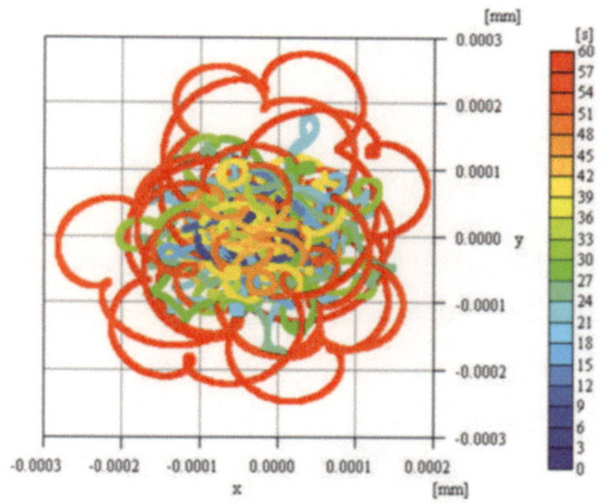

(b) Trajectory in y–x plane

Figure 19. Trajectories of the rotor magnet centre for 60 s starting from the restoring centre (0, 0, 99.5)

In contrast, the magnetic top, starting at the point 1 mm apart in z direction from the restoring centre, levitates in the range of ±0.002 mm in both x and y directions and ±1 mm in z direction around the restoring centre, as shown in Figure 18(b). The maximum tilt angle is 0.001165°.

Figure 19 shows the simulated trajectory of the centre of the rotor magnet, rotating at 1,380 rpm for 60 s starting from the restoring centre (0, 0, 99.5). In this case, a magnetic top levitates in the area of ±0.0003 mm in x–y plane and +0.016 mm/−0.003 mm in z direction. These results show that a rotating magnetic top can maintain levitation within several micrometres displacements in both radial and vertical directions.

Thus, the magnetic top has the ability to function as an entirely passive magnetic bearing [7].

4.4. Effects of the air drag force

In the previous analysis, the aerodynamic effects were neglected to simplify the discussion. If a magnetic top is rotating in air, rotating speed of the top will decay because of the pneumatic resistance acting on the surfaces of the top. In actual, the experiments showed that the rotating speed of the magnetic top decreases as time passes and the attitude of the top changes to a larger precession that leads it to fall down in a few minutes. Because there are no conducting materials in the magnetic top, there is no electrodynamic drag force caused by eddy currents. Hence, the aerodynamic drag force can be considered as the main reason for the decreasing rotation speed. In this section, some simulations are performed based on the equations of motion considering the aerodynamic drag force.

The aerodynamic effects to the behaviour of a rotating magnetic top is estimated as the pneumatic resistance acting on the outer side surface of the magnetic top. Here, the aerodynamic drag effects caused by the pole surfaces of the magnetic top are neglected. The following expressions are added to estimate the aerodynamic effects to the rotating speed of the magnetic top:

$$\omega_{n+1} = \omega_n - \frac{F_d}{I_{Top}} dt \tag{15}$$

$$F_d = \frac{1}{2} \rho C_d A_r v_r^2 r_{ro} \tag{16}$$

where $\rho = 1.225$ kg/m³ is the density of air, C_D is the coefficient of pneumatic resistance, $A_r = 2\pi r_{ro} h$ is the area of the outer side surface of the top and $v_r = r_{ro}\omega$ is the velocity of the outer side surface of the rotating top [8].

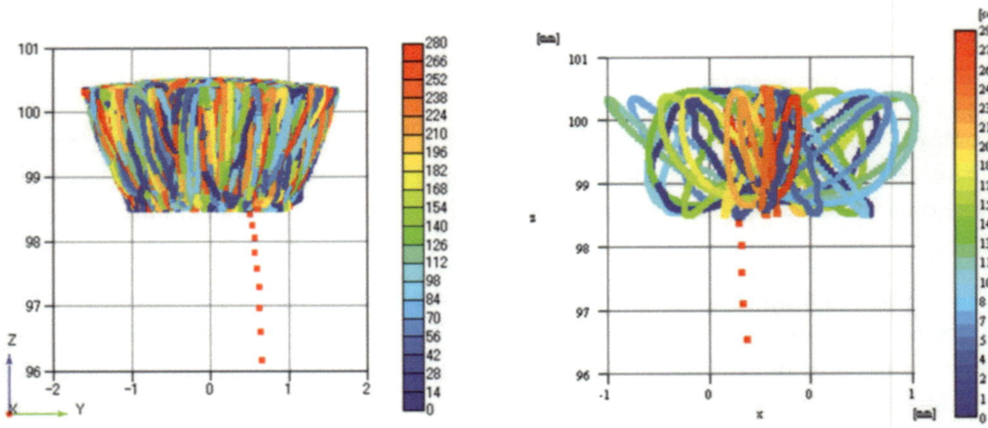

(a) $C_D = 0.5$ (b) $C_D = 5.0$

Figure 20. Simulated trajectory of the levitating magnet centre, initial rotating speed is 1380 rpm.

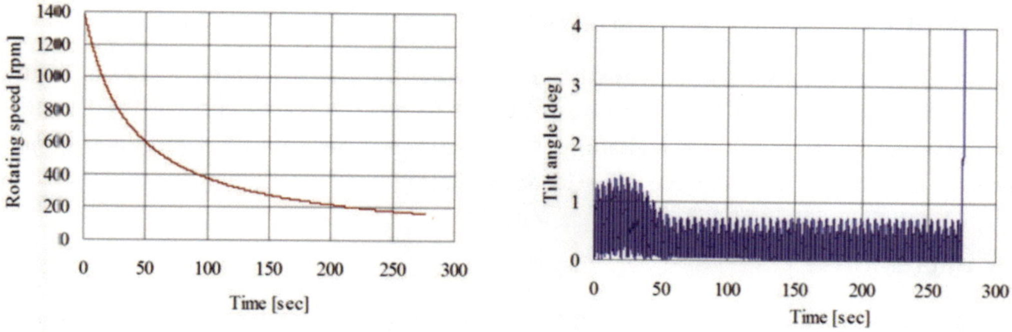

(a) Rotating speed of a magnetic top vs. time. (b) Tilt angle of a magnetic top vs. time.

Figure 21. Time dependence of rotating speed and tilt angle of a levitating magnetic top.

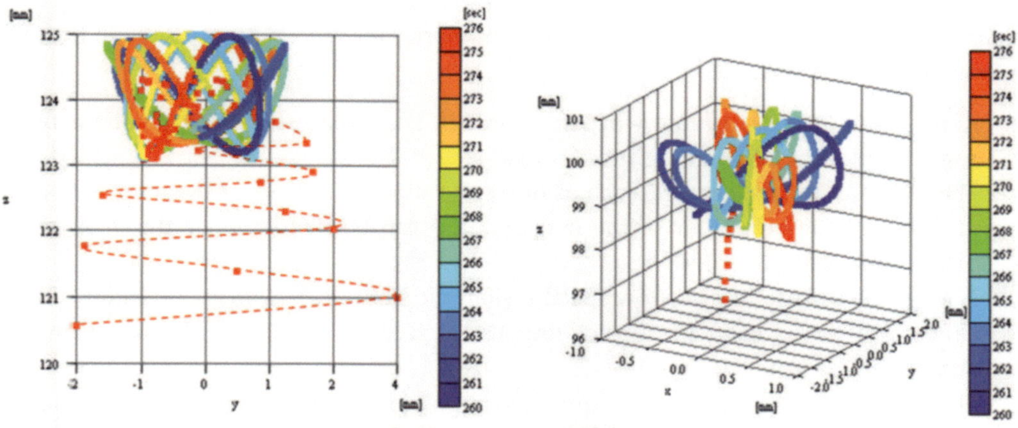

(a) Trajectory of the shaft head in the final 15 s. (b) Trajectory of the centre of the rotor magnet.

Figure 22. Trajectories of a levitating magnetic top started at 1380 rpm and fall down at 275 s after start.

Figure 20 shows the simulated trajectory of the centre of the levitating magnet starting from 1 mm apart in both x and z directions from the restoring centre. Initial rotation speed is set to be 1,380 rpm. The coefficient of pneumatic resistance C_D is set to be 0.5 for Figure 20(a) and 5.0 for Figure 20(b). Figure 20 shows that the magnetic top can levitate for 275 s or 28.4 s, if the coefficient of pneumatic resistance C_D is 0.5 or 5.0, respectively. Experiments showed that the magnetic top can be levitated for 3–4 min. Hence, in this study, the coefficient of pneumatic resitance C_D is assumed to be 0.5.

Figure 21 shows the time dependence of the rotation speed and the tilt angle of the levitating magnetic top. Figure 21(a) shows that the magnetic top started at 1,380 rpm and maintained levitation till 166 rpm at 275 s.

Figure 21(b) shows the time dependence of the tilt angle of the rotor magnet with respect to z-axis. This figure shows that the tilt angle varies within 1.4° while the top is levitating and indicates that precession is needed to maintain levitation for a magnetic top.

Figure 22 shows the trajectories of the magnetic top at the last 15 s of its levitation. Figures 22(a) and (b) demonstrate the trajectories of the shaft head and the centre of the levitating rotor magnet in its final 15 s levitation. These figures show that the precession quickly becomes larger once the magnetic top exits the levitating area.

5. Conclusions

A magnetic top levitates by itself, without any active control system, so long as it rotates in a certain speed range. The authors propose a simple and intuitive analysing method to predict characteristics of the magnetic top.

The quasi-three-dimensional static analysis, considering shapes and layout of the ring-shaped rotor and stator magnets, is used to explain the principle of levitation and obtain the preliminary design parameters of the rotor and stator magnets. The behaviour of the magnetic top is also investigated by dynamic simulations based on the three-dimensional equations of motion considering the moment of inertia for the rotating magnetic top. The following results are obtained:

1. A magnetic top can levitate when it rotates in precession mode with a slight tilting angle and in a certain rotating speed range.
2. The effects of the parameters, such as outer and inner diameters of the rotor and stator magnets, to the behaviour of the magnetic top can be discussed by the quasi-three-dimensional static analysis.
3. A magnetic top rotating at a low speed levitates in both precession and nutation modes. On the contrary, a magnetic top rotating at a high speed levitates with precession, and nutation mode is not observed.
4. The lowest rotation speed is determined to maintain the attitude of the rotating magnet using its mechanical inertia. The maximum rotation speed is limited by the centrifugal force that increases the tilt angle of the shaft of the magnetic top.
5. A magnetic top starting its rotation at the restoring centre will maintain its position with the accuracy of several micrometres.
6. It is difficult to repeat experiments of a magnetic top in the same conditions. However, the proposed analytical results showed good accordance with the experimental data observed using a digital video camera.

A magnetic top may be used as a rotating demonstration model in which some swaying motion can be permitted such as in toys or other relaxation items. When a magnetic top is used in commercial system, a rotor should be rotated by some non-contact drive mechanism such as electric motor or air turbine, etc. Furthermore, touch down bearing should be equipped to suspend a rotor while rotating speed is out of operating range. Fundamental requirements to design a magnetic bearing based on the principle of a magnetic top will be rotor weight and rotation speed. A rotor shaft should be designed considering mechanical requirements such as torque.

In the experimantal model, because ferrite magnets are used for the rotor and stator magnets, the restoring forces are very small for commercial applications. However, if rare earth permanent magnets and rigid suspension devices are used, sufficient restoring forces may be expected to be generated for use as a commercial passive magnetic bearing.

Author details

Teruo Azukuzawa
Japan Transport Safety Board, Tokyo, Japan

Shigehiro Yamamoto
Graduate School of Maritime Sciences, Kobe University, Kobe, Japan

Acknowledgement

The authors thank Mr. Makoto Matsumoto, former student of the Graduate school of Natural Science and Technology, Kobe University, for his efforts in establishing simulation tools for analysing the dynamic behaviour of a magnetic top.

6. References

[1] The Magnetic Levitation Technical Committee of the IEEJ (1993) Magnetic Suspension Technology - Magnetic Levitation Systems and Magnetic Bearings. Corona Publishing Co. Ltd., in Japanese.

[2] Matsumoto M, Azukizawa T (2004) Characteristics of Magnetic Guidance Force Between Coaxial Ring Shaped Permanent Magnets. IEEJ Technical Meetings on Linear Drives, LD-04-93, in Japanese.

[3] Miguel A.S. et al. (2005) Numerical integration for the dynamics of the heavy magnetic top. Physics Letters A 335 235–244.

[4] Ebihara D , Suzuki T (1988) The Repulsive Characteristics of the PM Type Magnetic Levitation Devices. Trans. IEEJ, Vol.108-D, No.5, 455-461, in Japanese.

[5] Azukizawa T, Matsuo N (2007) Feasibility Study of a Magnetic Top As a Magnetic Bearing. Proc. of the 6th International Symposium on Linear Drives for Industrial Applications, LDIA 2007(CD), 138.

[6] Azukizawa T, Yamamoto S, Matsuo N (2008) Feasibility Study of a Passive Magnetic Bearing Using the Ring Shaped Permanent Magnets. IEEE Trans. on Mag., Vol.44, No.11, 4277-4280.

[7] Azukizawa T, Yamamoto S, Makino H (2008) Effects of the System Parameters to the Behavior of a Magnetic Top. MAGLEV08, No. 63.

[8] Azukizawa T, Yamamoto S, Makino H (2009) Analysis of Dynamic Behavior of a Magnetic Top Considering Aerodynamic Drag Force. LDIA2009

Chapter 7

Theoretical and Experimental Investigations of Dynamics of the Flexible Rotor with an Additional Active Magnetic Bearing

Dorota Kozanecka

Additional information is available at the end of the chapter

http://dx.doi.org/10.5772/51113

1. Introduction

The conversion of energy in rotating machines is accompanied by phenomena that cause the additional dissipation of energy, affect technological processes, lower the endurance of machine elements and sometimes cause damages. Among these phenomena there are synchronous vibrations and self-excited lateral vibrations of the rotor and momentary pitches of amplitude of the rotor-bearings-foundation system vibrations.

Among the machines used in power engineering, rotary machines (compressors, pumps, blowers and turbines) are the ones used most often. Recent studies have put emphasis on the dynamics of rotating machines. This allows us to minimise the vibrations of the machine both during the period of its construction and at the time of its operation. The growing demand for reliability of rotating systems (i.e. API code) makes it necessary to specify their vibration parameters (critical frequencies, separation margins, amplitude of synchronous vibrations, permissible unbalance, etc.).

The search for new solutions of bearing systems in modern rotary machines that have to satisfy special performance demands has resulted in interest in rotor active magnetic suspension systems. The application of magnetic bearings as a system of shaft suspensions gives supplementary, unparalleled in classical solutions, diagnostic capabilities [10-14]. There are, however, also high requirements concerning the control system of the shaft position. New solutions in bearing systems have been more and more frequently applied in modern rotating machines. These include magnetic bearings that enable the active control of rotor vibrations.

A machine with a rotor supported in magnetic bearings allows for:

- enhancement of the general efficiency due to the lack of a mechanical contact between the journal and the bush, and thus, a low level of power loss in comparison with classical slide bearings characterised by the same load carrying ability, and due to elimination of an oil system and seals connected with it,
- maintenance of absolute cleanness of the working medium and a hermetic structure of the whole machine without the rotating end-piece of the shaft outside the casing in the case of an application of a high-speed electric engine integrated with the shaft,
- operation under high rotational frequencies, in a wide range of temperatures (from –160^0 to 250^{0}C) in chemically aggressive environments or in vacuum.

An active magnetic bearing system is a qualitatively different technology in comparison with classical solutions and requires the co-operation of specialists from two branches of technology, as it is a combination of a mechanical system with an electronic automatic control system, which controls this mechanical system [1,10].

A scheme in Figure 1 presents an active magnetic bearing as an automatic control system for one of the control axes - *y*. The voltage signal from the displacement transducer U_{DT} is conditioned. A change in the position of the journal with respect to the reference position (U_{Ref} -U_{DT}) activates the control current flowing through the bearing bush winding (electromagnet) in the electronic control system. This control results in a change in the electromagnet forces F_m that brings the journal to the assigned position.

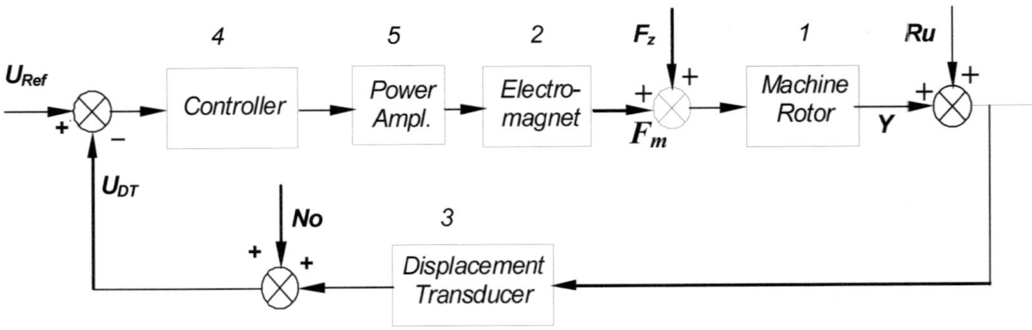

Figure 1. Scheme of the active magnetic bearing as an automatic control system (one of the control axes - *y*).

A proper value of the current is generated by the power amplifier on the basis of the signal provided by the controller according to the assumed control algorithm as a function of the present position *y* of the journal, measured by means of displacement transducers (Figures 1, 2).

Apart from disturbances connected with forces coming from e.g. unbalancing F_Z and static loading forces F_{stat}, the bearing system is affected by accidental interference *No* (noise) introduced by displacement transducers and interference following from heterogeneity of the measuring path of the runout type *Ru*. A level of these disturbances has a very significant influence from the viewpoint of stable operation of the system. A structure and algorithm of the applied controller have to ensure the system resistance to their effects.

An active magnetic bearing comprises two distinct components:

- a bearing itself, and
- an electronic control system.

A radial bearing comprises a rotor on which ferromagnetic laminations are fitted. The rotor is held in position by four electromagnets placed equally around the rotor, normally at 45 degrees to the vertical axis (Figure 2). The position of rotor (1) is monitored constantly by sensors (3) that detect any deviation from the nominal position. Any such deviation results in a signal which, by means of electronic control system (4, 5), is compared to the reference value. The resulting signal increases or decreases the current flowing to electromagnets (2) and thus returns the rotor to its nominal position. Each electromagnet produces an attractive force acting on the rotor.

An increase in high-speed turbomachinery applications has led to a significant progress in the research of rotor dynamics in the last decade. The evaluation of the dynamic stability and response to unbalance has become a standard calculation procedure for all new turbomachinery designs.

In order to apply magnetic bearings to suspend rotors of real machines, proper design methods that account for a special character of their operation and that are adapted to the requirements the bearings are to satisfy are needed. There is a desirable tendency in rotating machines to design rigid rotors because of a relatively low level of vibrations in their whole operating range which is observed. This aim can also be achieved by using flexible rotors with auxiliary magnetic bearings whose adjustable characteristics may be used to exceed safely the critical frequency of the rotor - bearing system with low amplitude of vibrations.

This work presents an idea of maintaining a low level of vibrations in the whole operating range of the rotating system of the flexible rotor (including critical speeds) by using an additional active magnetic bearing where the biggest anticipated dynamic deflection of the rotor occurs. What is achieved is the momentary operation of an additional magnetic bearing in the rotating system before a high level of the amplitude, which corresponds to the critical frequency of lateral vibrations. This causes a qualitative change in the dynamic properties of the system. It also allows one to reach the nominal revolutions of the machine without the dangerous effects connected with exceeding the critical frequencies of the flexible rotor.

The realisation of the presented idea for the real rotating system requires a preparation of the theoretical and experimental methods of investigations and numerical simulations for the model of active magnetic suspension.

In this work the theoretical and experimental investigations of the digitally controlled active magnetic bearing and the theoretical and experimental investigations the flexible rotor dynamics carried out on the test stands with an auxiliary active magnetic bearing with a digital control system are discussed [5,6,9].

2. Digitally controlled magnetic bearing

Active magnetic suspension systems of machine rotors being built at present are equipped with digital control systems. Apart from a possibility of implementation of complex control algorithms, they provide also wide diagnostic possibilities resulting from an application of measurement techniques at different stages of the system design. Control systems of bearing responses decide about dynamic properties of the rotating system. Digital controllers allow for, e.g. a change in bearing dynamic properties during motion in different modes of the machine operation.

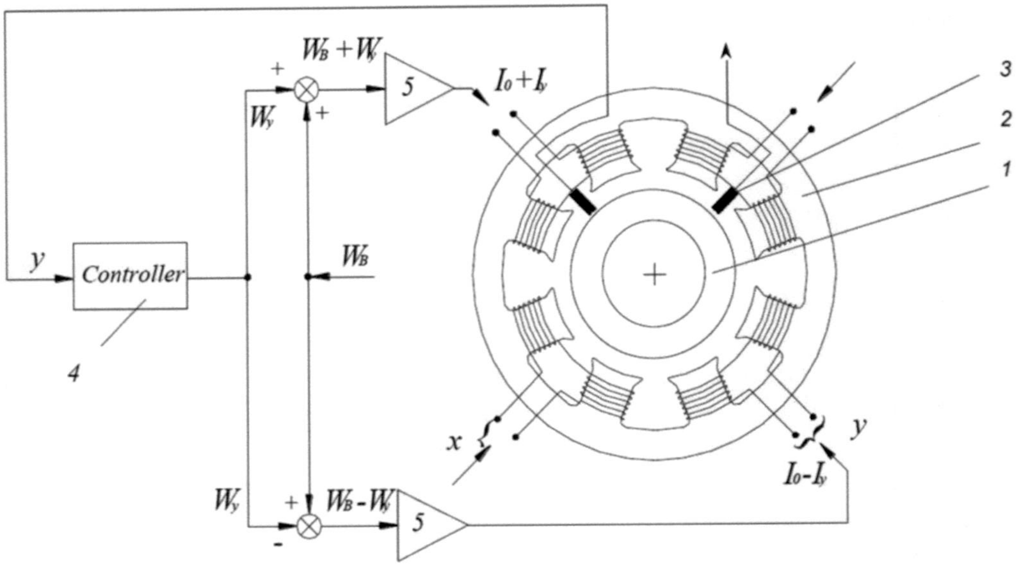

Figure 2. System of the digitally controlled magnetic bearing (one of the control axes - y).

The research on active magnetic bearing technology, including works on digital controllers and algorithms, actuators and magnetic bearing-rotor system dynamics, has been carried out for several years in the Institute of Turbomachinery of the Technical University of Łodz [4-9]. The mechanical structure of the built active magnetic bearing, consists of a journal and a bush with four pairs of electromagnets placed equally around the rotor. The position of ferromagnetic journal (1) with respect to bush (2) is controlled by means of eddy-current displacement transducers (3) made by *Bently Nevada Corporation*, with the diameter $d=8$ *mm* and the static sensitivity *7.870 V/mm*. They are mounted on two control axes x, y that are perpendicular with respect to each other and displaced by the angle 45^0 with respect to the journal axis. The control axes interact with respective pairs of electromagnets (Figure 2).

Each pair of the bush electromagnets of the journal bearing interacts with a digitally controlled power amplifier with a variable pulse width *PWM* (5). The control pulse-width modulation W_y and W_x is counted by controller (4) on the basis of measurements of the position y and x of the journal, respectively. The control current that supplies the windings of individual electromagnets is an explicit function of this modulation. For the journal magnetic bearing, the differential control is used and it requires a generation of the bias current I_0 . To obtain the bias current I_0 for the designed bearing, the pulse-width modulation W_B, which determines its operating point on the characteristics, is used.

3. Numerical model of the magnetic bearing system

The basic assumption while developing the numerical model of the bearing system was to offer a tool that allows for tuning the parameters of its controller and for carrying out the investigations of the designed system dynamics in a wide range [4,14]. The condition to be met during the realisation of the idea of the numerical simulation of the bearing was to reproduce the algorithm of operation of the real bearing actuating system structure, and the measurement and control elements applied in this real system in the model. The fulfilment of this requirement has guaranteed the correctness of the operation of the model and feasibility of its design.

A general scheme of the system that has been employed in order to develop the numerical model is presented in Figure 3a. In this model, a motion of the mass *m* concentrated in the geometrical center of the journal is analyzed.

A controller used in the system controls both the axes *x* and *y*. For each axis, differential control with a programmed value of the *pulse-width modulation* of the so-called W_B base has been applied. Figure 3b shows system of the position of the journal with respect to the bearing bush: *EM1, EM2* – bush electromagnets interacting with the axis *y*, *EM4, EM3* – bush electromagnets interacting with the axis *x*, F_Z – rotating vector of residual unbalancing, F_{mx} – electromagnet force acting along the axis *x*, F_{my} – electromagnet force acting along the axis *y*, $F_g + F_{stat}$ – forces of gravity and static load.

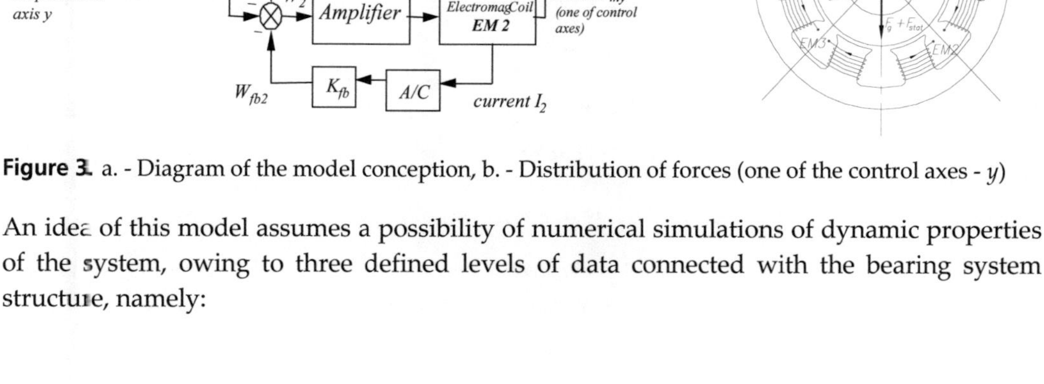

Figure 3. a. - Diagram of the model conception, b. - Distribution of forces (one of the control axes - *y*)

An idea of this model assumes a possibility of numerical simulations of dynamic properties of the system, owing to three defined levels of data connected with the bearing system structure, namely:

- physical parameters of the bearing: mass supported in the bearing, radial clearance of the magnetic bearing, initial position of the journal with respect to the bush;
- parameters of the controller structure and the control system structure: boundary values of the pulse-width modulation W_{min} , W_{max}, and W_B which determines the operating point on the characteristics, feedback coefficient K_{fb}
- parameters of the actuating systems: frequency of control pulses f_{PWM} , power amplifier supply voltage U_Z , inductance of the actuating system L, resistance of the actuating system R, value of the bearing constant K.

The model gives also a possibility of analysis of permissible levels of disturbances F_Z, Ru, No that provide a proper margin of the bearing system stability for a given type of the controller. Thus, in a sense, it constitutes the synthesis of a robust controller.

The numerical procedures representing actual characteristics of the actuator were developed and verified, and then applied in the simulation model of the bearing. This allows for modelling the magnetic bearing system quickly and accurately. The numerical simulation of the active magnetic bearings system by means of the Hewlett-Packard HPVEE software was elaborated.

In the designed digital control system of a journal active magnetic bearing, the suitable software that allows for investigations of the *pulse-width modulation - control current* characteristics that determine the bearing system properties while its actuators are built, has been applied.

3.1. Numerical characteristics pulse-width modulation - control current

Each pair of the bush electromagnets of the built journal bearing interacts with a digitally controlled power amplifier with a variable pulse width *PWM*. The control pulse-width modulation W is counted by the controller on the basis of measurements of the position of the journal with respect to the bush. The control current that supplies the windings of individual electromagnets is an explicit function of this modulation [4,9].

Power transistors and discharging diodes have been used to build the amplifiers (Figure 4). They are characterized by specified values of operating parameters. The properties of transistors and diodes, as well as the winding inductance and resistance of a given pair of poles, determine the parameters of the whole bearing actuator path, in which control current is generated.

The dynamics of changes of the current in windings is strictly dependent on the force generated by electromagnets in the gap between the journal and the bush and it affects directly the dynamics of the mass suspended in the bearing. The knowledge of actual characteristics of the actuator paths is an important factor in designing the structure and the algorithm of the controller, whose task is to ensure stable operation of the bearing.

According to these assumptions, a development of the program that allows for numerical simulation of the theoretical equations describing the phenomena occurring in the actuator electric circuit versus time (i.e., in subsequent periods of control pulses) for nominal parameters of the power amplifier elements, electromagnet windings, supply voltage and control frequency, was required.

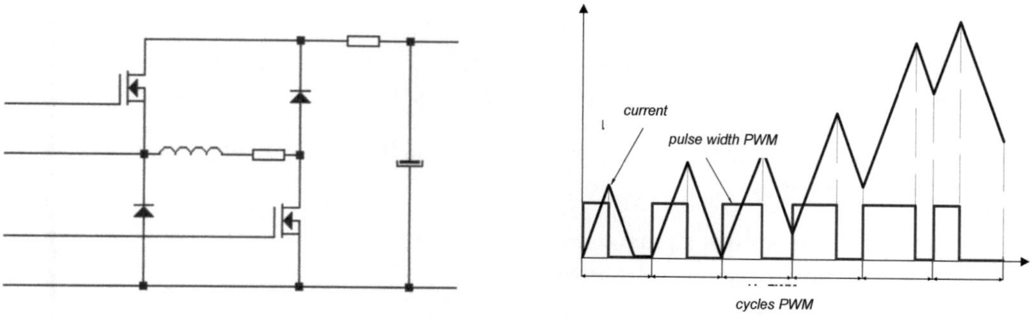

Figure 4. Power amplifier circuit and an idea of current changes in the bearing

An operation of the actuator was analyzed. Each cycle of the control pulse *PWM* of a given frequency f_{PWM} and a pulse-width modulation W forces two modes of the amplifier operation, namely:

- charging, whose duration is equal to $t_{char} = W / f_{PWM}$. Elements of the amplifier circuit operate then in the conduction mode,
- discharging, whose duration is equal to $t_{dischar} = (1 - W) / f_{PWM}$. Elements of the amplifier circuit operate in the lockout mode.

For a given pulse-width modulation, a value of the control current $I_{av\ i}$ that generates the bearing magnetic force results from its averaged value in the *PWM* cycle and has been expressed by relation (1):

$$I_{avi} = W_i \frac{I_{0i} + I_{\max i}}{2} + (1 - W_i) X_i \frac{I_{\max i} + I_{0i+1}}{2} \tag{1}$$

$I_{0\,i}$ – initial current for the i – *th* cycle of the amplifier charging.
$I_{0\,i+1}$ – initial current for the $i+1$ – *th* cycle of the amplifier charging.

The coefficient X_i occurring in formula (1) determines a ratio of the discharging time for a given cycle to the time during which the power amplifier electronic elements cause that the control current value diminishes to zero. In the cycles in which the current discharges to $I_{min\,i} = I_{0\,i+1}$, the coefficient $X_i = 1$.

Individual operation sequences of the digitally controlled pulse power amplifier and the windings powered by it have to be analyzed. Numerically simulated sequences of the ideal actuator operation enable one to generate model characteristics, which are the basis for evaluation of an influence of changes in values of individual parameters of the actual system on their shape.

Numerical calculation procedures that allow for the generation of instantaneous and mean time histories of control currents in the assumed *PWM* cycles, for the following actuator parameters: amplifier supply voltage U_z, inductance of the bearing bush winding L, resistance of the power amplifier circuit R, frequency of control pulses f_{PWM} as a function of values of the pulse-width modulation coefficient W of the control pulse, have been developed.

In Figure 4 a and b exemplary *pulse-width modulation - control current* characteristics of the bearing actuator system calculated for selected values of its parameters are shown.

Nominal, experimentally identified values of the parameters of the actuators assumed in the calculations were as follows:

- inductance of the electromagnet winding $L = 80mH$,
- resistance of the power amplifier circuit $R = 3.1\Omega$.
- supply voltage of the power amplifier $U_z = 80V$,
- frequency $f_{PWM}=1667Hz \Rightarrow T_{PWM}= 600\mu s$.

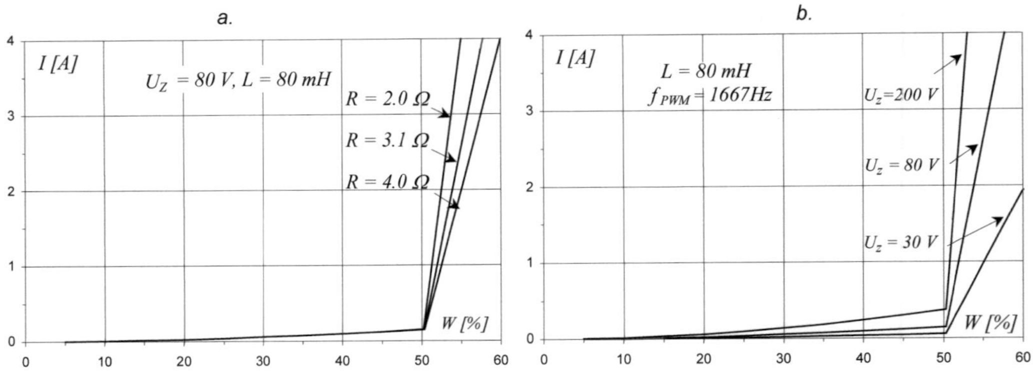

Figure 5. *Pulse-width modulation – control current* characteristics of the bearing actuator
a. - as a function of the amplifier circuit resistance, b.- as a function of the power amplifier supply voltage

The characteristics were calculated for experimentally identified parameters of the actuator systems in order to enable their verification for a real object at each stage of the calculations.

To show a tendency of changes, the characteristics generated for bigger and smaller parameters than the nominal ones have been presented in the figures as well. This allows one to forecast and evaluate a possibility of introducing changes in the values of parameters of the designed bearing system actuator. It is of great practical importance.

From the characteristics shown in Figure 5a and b, it follows that for the structure of the power amplifier under investigation and for the real parameters of bush windings, the actuator has to operate under pulse-width modulations W>50% in order to ensure a suitable load capacity.

For a pulse-width modulation *W>50%*, the dynamics of changes in the control current depends on resistance in the power amplifier circuit. The total value of resistance includes electromagnet winding resistance and dynamic resistance of the amplifier electronic elements in subsequent cycles of its pulsating operation, i.e. during charging and discharging (Figure 5a). An increase in the dynamics of the current changes in the bearing system under consideration, i.e. for given, actual parameters of the bush windings and the power amplifier, can be achieved through an increase in the value of the power amplifier supply voltage *Uz* (Figure 5b).

3.2. Experimental characteristics pulse-width modulation - control current

In order to identify the properties of each actuator path, suitable algorithms, software for a microprocessor of the measurement-control card and a host computer of the bearing system have had to be prepared. Independent auxiliary procedures that allow for experimental testing the actuator paths of the bearing have been developed. These procedures form a program for operation of the measurement-control card, which is the main element of the bearing digital control system [11].

An active magnetic bearing system comprises actuators, which are realised in the form of windings of individual pairs of bush electromagnets that interact with digitally controlled power amplifiers. A scatter of real values of electric parameters of the winding for individual pairs of electromagnets, to which a technological scatter of the amplifier structure elements is added, will introduce the asymmetry of properties for the control axes.

From the point of view of simplification of the structure designing and the controller parameter tuning in order to obtain the system stable operation, individual bearing actuators should be characterised by the symmetry of properties. A fulfilment of this condition will allow for making the mechanical characteristics of the bearing system independent of the actuator properties for each axis.

Control systems of the bearing response are a very important part of a machine with active magnetic bearings because they decide about the dynamic characteristics of the rotor. In order to design and build a control system fulfilling the requirements of the object under investigation, the knowledge of its real model is needed.

An active journal magnetic bearing with a digital control system has been built according to the presented idea. A new concept of control systems has been elaborated. It allows one to program the control of the bearing actuator characteristics. To achieve this goal, a programming procedure for the control processor of the bearing has been developed. The programmed testing procedure of an actuator allows one to reach experimental characteristics for its various parameters (frequency and width *PWM*, power supply, air gap between the rotor and the stator) for each control axis [5].

The experimentally determined *pulse-width modulation W - control current I* characteristics show differences in the bearing properties for two control axes x and y (Figure 6a). As a result, they render the asymmetry of mechanical properties of the system for these axes when the feedback loop is closed and the bearing control system is turned on.

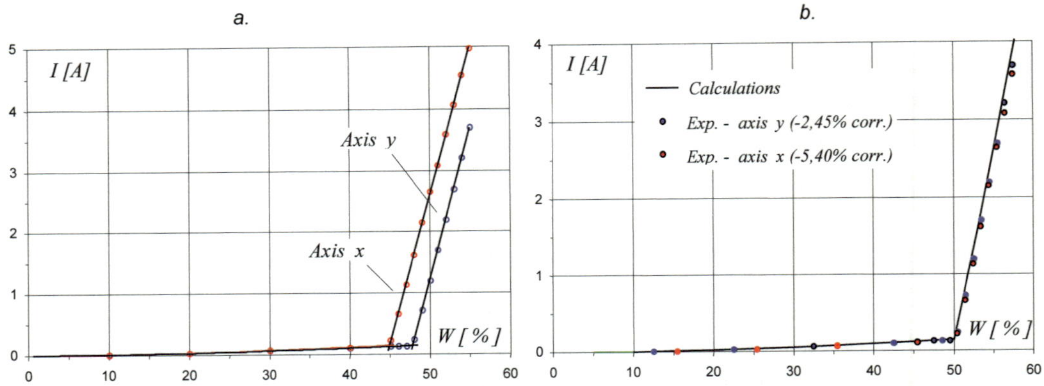

Figure 6. a. - Experimental *pulse-width modulation W - control current I* characteristics of the bearing actuator systems before programmed correction, b. - Theoretical and experimental characteristics after programmed correction.

For the given parameters of a real bearing system, there is always a scattering of properties of the actuating system for each pair of electromagnets. The developed procedure enables one to tune these characteristics in order to obtain the symmetry of operation of actuating systems for each pair of electromagnets and for each bearing control axis *y* and *x*.

An alternative to compensation for the asymmetry of bearing actuator properties by designing a proper structure and algorithm of the controller is to develop a method of its programmed correction. On the assumption that the source of the identified asymmetry lies in a scatter of actuator parameters, a correction method that leads to minimisation of the effect of the scatter of their characteristics for each bearing control axis has been proposed.

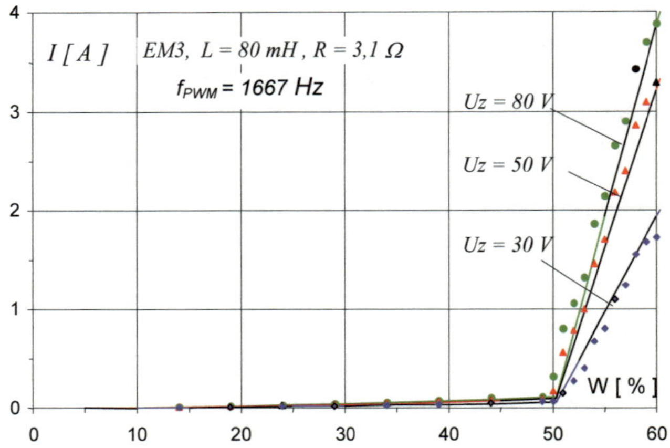

Figure 7. Theoretical (solid line) and experimental (dots) *pulse-width modulation - control current* characteristics of the bearing actuator systems after the programmed correction for various values of the power amplifier supply voltage

From the *pulse-width modulation - control current* characteristics that were experimentally determined (Figure 6a), it follows that the dynamics of the current changes increases already at the pulse-width modulations W<50% in an actual actuator. These phenomena are connected with the structure of control systems, power amplifier transistor gates and they need a necessary correction. The theoretical characteristics are used to evaluate a scatter of properties of real actuator paths of the bearing. This evaluation allows one to introduce necessary programmed corrections of characteristics of real actuator systems to achieve the convergence between the characteristics determined experimentally and those calculated numerically [6].

This correction results in obtaining the symmetry of properties of individual actuators of the designed bearing which interact with the bush electromagnet windings *EM1, EM2, EM3, EM4* (Figure 6.b).

After this correction, *pulse-width modulation - control current* characteristics of actuator paths of the bearing have been determined experimentally. The results of experimental investigations and numerical simulations have exhibited a good convergence. Figure 7 shows these characteristics for the bearing actuator path interacting with the axis x - *EM3* for different values of the power amplifier supply voltage.

The proposed way of the programmed correction of the characteristics has been verified experimentally and carried out for a real bearing structure [4,5,6].

3.3. Numerical calculations and verifications

In the analysis of model investigation results, the same quantities have been chosen as those recorded in a real magnetic bearing system, namely: pulse-width modulation, displacement, static equilibrium position, journal trajectory, phase portrait, Bode's plot, etc. It has enabled their direct verification with the experiment. The verification has been carried out on the test stand with the built radial bearing, whose part is presented in Figure 8.

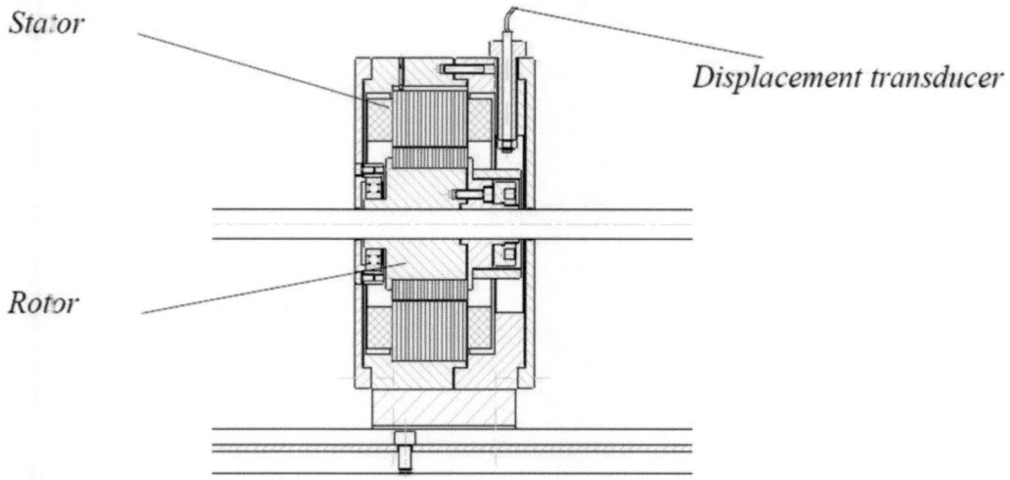

Figure 8. Structure of the built radial bearing

The investigations of the bearing dynamics comprise three basic modes of its operation:

- suspension of the journal in the bush,
- start-up,
- operation under constant rotational frequency.

Some selected test results for the first mode of the system operation are presented.

The nominal values of parameters that have been assumed in the model calculations are as follows:

- power amplifier supply voltage $U_Z = 80V$
- frequency of control pulses $f_{PWM} = 1667Hz \Rightarrow T_{PWM} = 600\mu s$
- inductance of the electromagnet winding $L = 80mH$
- resistance of the power amplifier circuit, including the windings $R = 3.1\Omega$
- bearing constant $K = 2.4\ 10^{-5}\ Nm^2/A^2$

The bearing constant *K*, as well as the power amplifier circuit resistance and inductance, *R*, *L*, have been determined experimentally through suitable measurement procedures for the real structure of the bearing system, whose actuators operate at the above-mentioned values of the supply voltage U_Z and the frequency f_{PWM}.

In order to present the diagnostic capabilities of the model developed and to indicate a convergence of its operation with the real system, the suitably selected characteristics calculated numerically for the defined nominal parameters and those recorded for the real bearing system have been shown in Figures 9 - 13.

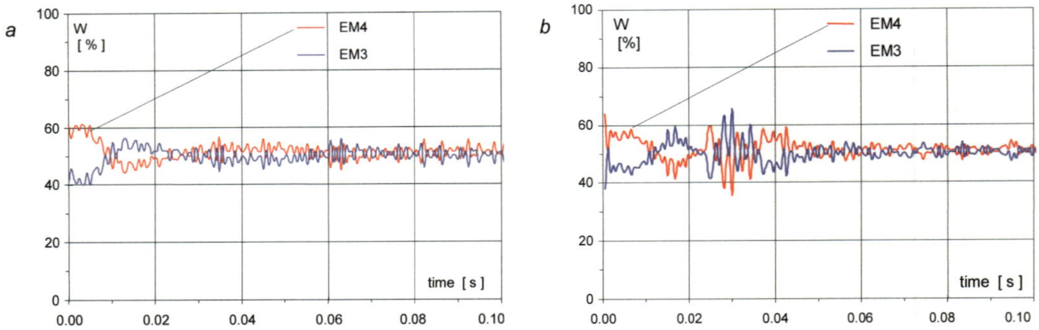

Figure 9. Changes in the pulse-width modulation during the suspension of the journal in the bush for the control axis x (electromagnets: top - *EM4* and bottom - *EM3*) – a. - numerical simulations, b.– experimental results

For one control axis during given sampling time, pulses with the pulse-width modulation *W* proportional to the signal produced by the controller are calculated on the basis of the displacement measurement of the journal with respect to the bush. After its comparison with the assigned pulse-width modulation that determines the bearing W_B operating point and the pulse-width modulation that corresponds to the value of the current flowing in the top and bottom electromagnet winding W_{fb}, the pulse-width modulation W_{fb} is realized through the application of an additional feedback loop with a possibility of programmed assigning the value of the feedback coefficient K_{fb}.

In Figs. 8 a and b changes in the pulse-width modulation as a function of time in the mode of suspension of the journal in the bush for the windings of the top electromagnet *EM4* and the bottom one *EM3* interacting with the axis x are shown. The data recording for the real bearing structure was carried out by means of the developed *PRDP* procedure. The presented calculation results and the recorded results refer to the same parameters of the controller. A good convergence of the results has been obtained, which confirms reliability of the model built.

On the basis of the calculated pulse-width modulations, the bearing actuators generate currents that control the windings of corresponding bush electromagnets. The experimentally verified procedures that represent the real actuator operation for the assigned parameters have been employed in the algorithm of the model operation.

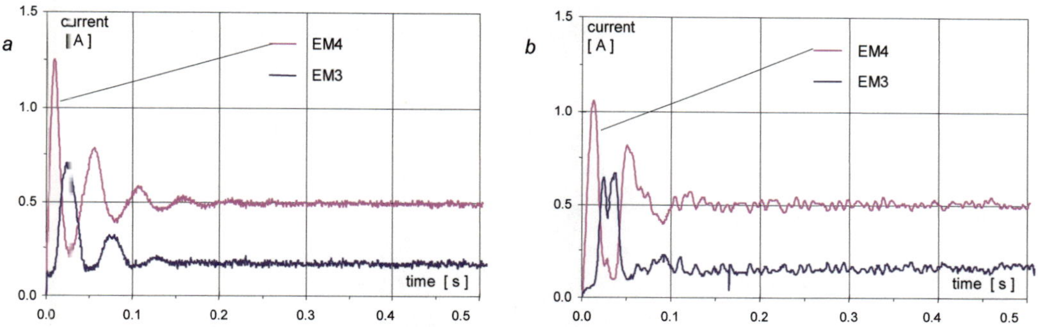

Figure 10. Currents in the windings of the top *EM4* and bottom *EM3* electromagnet - axis *x*
a. - numerical simulations, b. - experimental results

For the real structure, the values of correction, indispensable to obtain the symmetry of bearing properties, have been introduced for each pair of windings for both the control axes *x* and *y*.

In Figures 10 a and b changes in the current versus time are shown in the mode of suspension of the journal in the bush for the electromagnet top windings *EM4, EM1* and bottom windings *EM3, EM2,* interacting with the corresponding control axes *x* and *y*. The control currents in the bush top and bottom winding along each control axis are time-variable and are a function of the change in the journal position with respect to the bush until the equilibrium position is achieved.

Figures 11a and b show changes in the displacement of the magnetic bearing journal at the moment an active magnetic bearing is activated for both the control axes *x, y* for the real system and its simulation model, respectively.

In Figures 11 and 12 an effect of the bearing system operation in its first operation mode in the form of the calculated and recorded changes in the position of the journal with respect to the bush as a function of time is presented for both the control axes and for the orbit composed of these displacements.

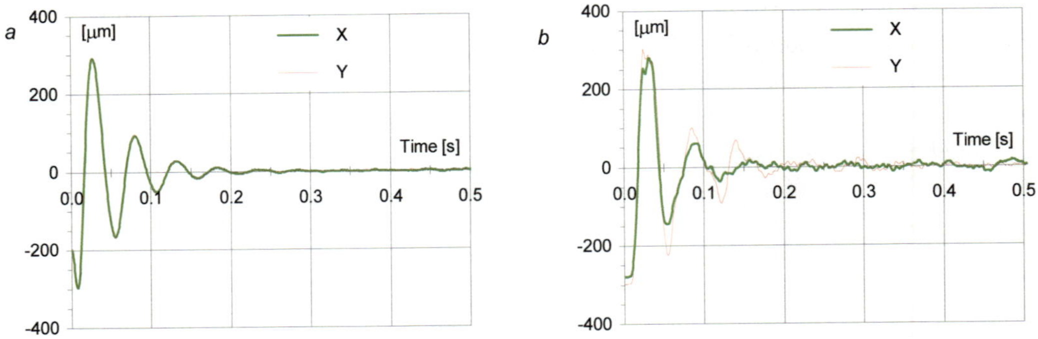

Figure 11. Displacement of the rotor a. - numerical simulation, b.- experimental results

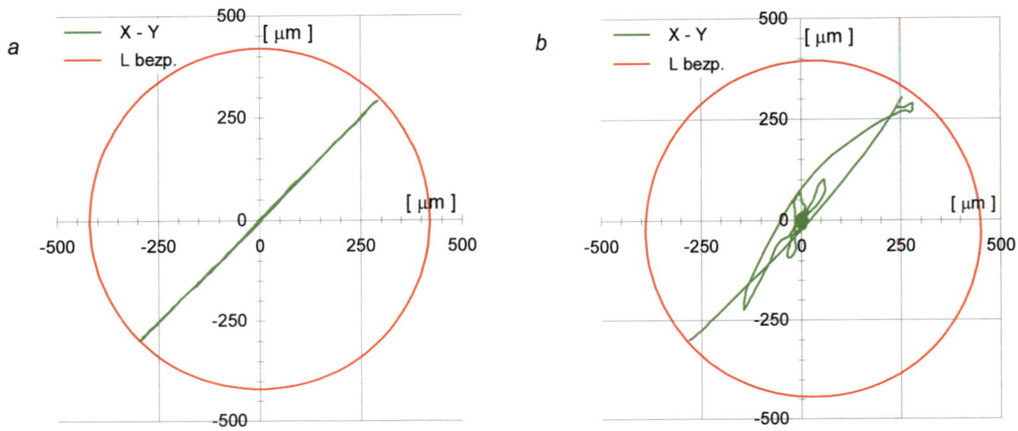

Figure 12. Orbit of the rotor a. - numerical simulations, b. - experimental results

At the ideal symmetry of the properties of the actuators for each control axis, which is assumed in the model, symmetrical time histories of the journal displacement for both the position control axes $x(t) = y(t)$ are obtained - Figure10 a. In the real system, slight differences in the time history of the journal displacement occur at the moment of its start-up in comparison with the model. These differences result from: non-ideal geometry of the journal, asymmetry of the arrangement of the position sensors with respect to the journal or asymmetry of the journal position with respect to the bush (Figures 11b, 12b).

The character of the changes in the journal position as well as the time that passes from the moment the control system is activated up to the moment in which stable suspension of the mass supported in the magnetic bearing is achieved are the same for the model and for the real structure under investigation.

In Figure 13 changes in the magnetic force generated in the bearing system during suspension of the journal in the bush versus time are shown. The characteristics is obtained on the basis of the simulations.

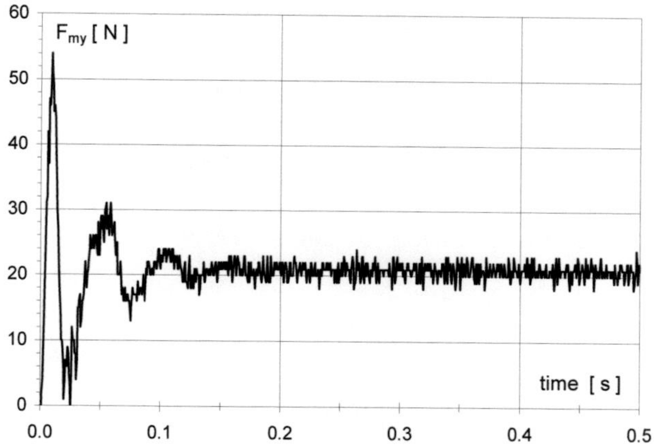

Figure 13. Magnetic force F_{ymag} generated in the bearing system - axis y

The agreement between the simulation investigations and the real bearing structure investigations has confirmed that the simulation model developed is a reliable tool that can be employed in designing a bearing system and in forecasting the tendencies of changes in its operation under the assigned level of disturbances and assigned excitations.

Their comparison indicates the feasibility of the simulation model in practical applications.

In order to apply simulation investigations in practice, their verification for an actual object is needed. The basic assumptions connected with the developed simulation model of an active magnetic bearing system are presented. The simulation and experimental investigations have been verified on the test stand for a digitally controlled journal active magnetic bearing. The results discussed confirm a convergence between the operation of the model and real bearing.

A reliable theoretical model that allows for analysis of the bearing dynamics under hypothetical, extreme loads reduces the designing time and enables one to minimize errors that can occur at the system prototype start-up.

The tool allows for testing different variants of the magnetic bearing system operation, which has been confirmed by the experimental investigations conducted in a wide range. A diagnostic capability of the non-linear numerical model of the magnetic bearing allows to develop a method for the identification of dynamical parameters: stiffness and damping [8,9].

4. Concept of the method for the identification of dynamic parameters

In the proposed measurement method, a dependence of the vector of the resultant magnetic force acting on the machine shaft as a function of the journal position in the magnetic radial bearing and the currents that flow through the windings of actuators (electromagnets) is employed. The magnetic bearing response vector is a sum of the forces generated by bearing electromagnets and alters in each control cycle *PWM* [5,9]. The value of the magnetic response component F_{Xmag} for one control axis is related to the measured mean values of the current controlling the electromagnets I_{XT}, I_{XB} in a given control period and the values of the magnetic gaps s_{XT}, s_{XB} (top – index *T*, bottom – index *B*). The values of the magnetic gaps are found on the basis of measurements of instantaneous positions of the journal with respect to the centre of the bush of the known clearance. The magnetic response component F_{Xmag} for the axis *X* is determined by the following relationship:

$$F_{X\,mag} = K_{XT}\frac{I_{XT}^{\;2}}{s_{XT}^{\;2}} - K_{XB}\frac{I_{XB}^{\;2}}{s_{XB}^{\;2}} \tag{2}$$

Equation (2) holds on the assumption that the linear dependence of the magnetic flux on the induction is maintained. It means that the bearing operates according to this part of the characteristics that is distant enough from the state of magnetic circuit saturation, when the induction does not exceed 50 % of the saturation induction for the core material. The value of the constant *K* depends on electromagnet design parameters and can be calculated theoretically [1]. However, in the actual design of the journal bearing, the constants K_{XT}, K_{XB} K_{YT}, K_{YB} can differ slightly for each pair of electromagnets of the bush. In order to increase the accuracy of the proposed measurement method, the constant values are verified experimentally for each electromagnet and their real values are taken into account in the calculations [8]. If the journal motion parameters are known and the magnetic response force is determined by an indirect method, it is possible to find the bearing dynamic parameters that relate the magnetic response force to the journal motion parameters [5,9].

An analysis of the system response to synchronous excitation points out to the fact that the bearing magnetic response force F_{xmag} F_{ymag} is proportional to the excitation force amplitude F_z in the whole range of frequency of rotations. It allows for the identification of equivalent dynamic coefficients of the bearing. In the proposed method a non-linear analysis of the simulation model in the assumed range of rotational frequency changes of the rotating mass of the bearing, for the excitation with the rotating vector of unbalance F_Z, is required. For each assumed value of the frequency of rotations, the value of unbalance F_Z is chosen in a such a way that the forced vibrations of the bearing rotating mass have the same value of amplitude in each control axes *x, y*. For the assumed frequency values *n*, the time histories of the following quantities are recorded:

- excitation unbalance force $F_Z(t)$,
- variable components of displacement of the rotor mass along the respective control axes $x_{ac}(t)$, $y_{ac}(t)$,
- velocity components of the rotor mass in the control axes $v_X(t)$, $v_Y(t)$,
- variable components of the magnetic response forces for the control axes $F_{xmag\,ac}(t)$, $F_{ymag\,ac}(t)$.

These time histories are non-linear, as no simplified assumptions have been introduced into the model for the relations between the force, current and displacement. In the final stage of the proposed method, a relation between the linear stiffness and damping coefficients and the bearing response force, is employed on the assumption of a lack of coupling between the bearing control axes. For one control axis:

$$F_{x\,lin} = K_{XX}\, x_{ac} + C_{XX}\, v_x \tag{3}$$

A difference between the non-linear magnetic bearing response force in a given axis $F_{xmag\,ac}$, known on the basis of model calculations and its linear form $F_{x\,ac\,lin}$ determined by the formula:

$$F_{x\,lin} - F_{xmag\,ac} = \Delta F \tag{4}$$

in such a way that $\Sigma \Delta F^2 = min$ is sought with the least squares method. Thus, the linearized coefficients of stiffness and damping K_{XX} *[N/m]*, C_{XX} *[Ns/m]* are obtained.

Changes in the non-linear magnetic response force $F_{mx\,ac}$, provided by the digital controller for a given control axis, which result from the numerical calculations, are approximated with the linearized harmonic time history $F_{x\,ac\,lin}$. Its values are determined by the dynamic stiffness K_{XX} and damping coefficients C_{XX} for a given control axis. An analogous situation refers to the coefficients K_{YY}, C_{YY} for the axis *Y*.

It is possible to use the proposed method for calculation of dynamic coefficients of the bearing when the developed simulation model of the bearing, whose operation is convergent with the operation of a real bearing system, is employed.

The calculations are conducted for stable bearing operation, where the journal position oscillates around the assumed point of equilibrium and the interactions between the control axes *X* and *Y* can be neglected for small displacements of the journal (Figure 14).

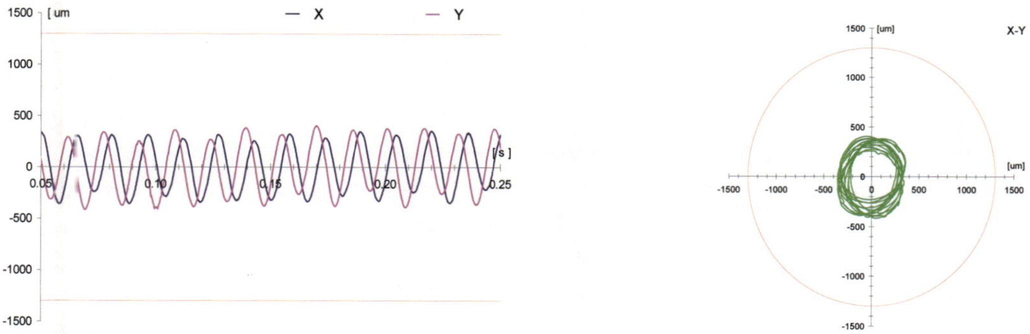

Figure 14. Displacements *X, Y* versus time and the orbit

Figure 15 presents a comparison between the measured magnetic response force F_{Xmag} and the theoretical function, which is a sum of the forces of stiffness and damping $F_{X\,lin} = K_{XX}\, x + C_{XX}\, V_X$.

The curve $F_{X\,lin}$ has been plotted on the basis of the measured journal displacement *x* (Figure 14) and the journal velocity V_X obtained through digital differentiation of the displacement and a selection of suitable values of the dynamic stiffness coefficients K_{XX} and the damping coefficients C_{XX} in such a way as to make the sum of squares of differences minimal for the selected part of the time history.

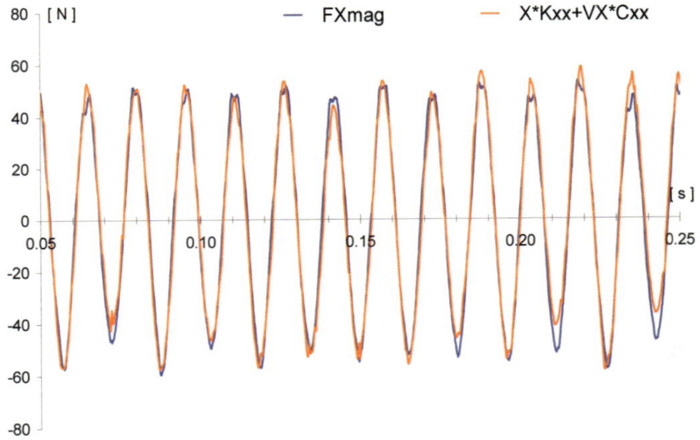

Figure 15. Measured magnetic response component along the control axis X - F_{Xmag} and its modelled time history $F_{X\,lin}$ with the identified dynamic coefficients K_{XX} , C_{XX}

In the method of identification of bearing dynamic coefficients, it is required that the theoretically calculated magnetic response force is the closest approximation of its function obtained in the measurements and that the share of synchronous components in the curves of displacement, current and magnetic response force is dominant [5,6,9].

5. Results of the investigations

The model test stand of the flexible power-transmission shaft is used to carry out the experiment whose aim is to verify the identification procedure of dynamic coefficients of the active magnetic bearing (Figure 16). The test rig consists of the horizontal shaft supported in two rolling bearings mounted at both ends. It is driven by an electric motor connected to the shaft through an elastic membrane coupling with smooth rotation control and fed by a frequency converter. An active magnetic bearing operates as an auxiliary bearing that modifies the dynamic properties of the shaft line. Between the magnetic bearing and the shaft right end, there is a rigid disk which allows one to mount balancing weights for the real structure. The mass of the rotating system is equal to *4.85 kg*, the shaft line length equals *1923 mm*. The test stand allows one to investigate the effects of the magnetic bearing on dynamic properties (vibration level, displacement and the coefficient of vibration amplification of subsequent critical frequencies) and to control vibrations of the long, flexible rotor. A kinematics exciter is fixed on the disk, on which masses of test unbalancing can be mounted. After introducing a selected program for magnetic bearing control, harmonic vibrations of the shaft of the frequency *(10, 20, 30, … ÷80) Hz* and the assigned amplitude were excited. For each frequency under analysis, the time histories of displacements and currents in the magnetic bearing, which were subject to respective calculation procedures, were recorded, and then the bearing dynamic parameters were estimated. To conduct the measurement and calculation procedures, a measurement system with *DBK 15* input systems made by *IOtech,* operating with a PC and employing the *Daq/112B* type *PCMCIA* measurement card of the resolution equal to *12 bites* and the maximum sampling frequency of *100kHz,* was applied.

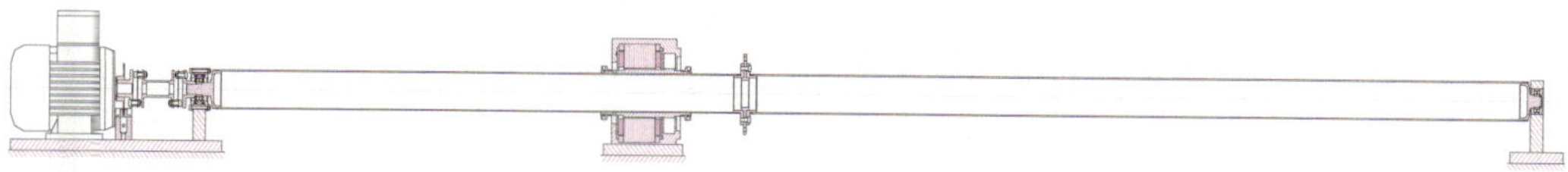

Figure 16. Configuration of the test rig

The voltage time histories corresponding to displacements (positions) of the journal along both the control axes *X,Y* were recorded *on-line* on respective inputs of the measurement-control module. These were two voltage signals *0-24V* from *Bently-Nevada* type *3300* eddy-current transducers of relative vibrations. The voltage time histories corresponding to currents flowing in electromagnets were measured and recorded. These were four voltage signals *0-5V* from current-voltage *LEM* type transducers.

The DaqView v.7.9.8 software was used for recording purposes. There were *4000* measurements made, at the sampling frequency of *8kHz/channel*. The results were stored in binary files of the data acquisition system, and then converted into text files. The programs for analysis of dynamics and identification procedures of bearing dynamic parameters, according to the methodology proposed, were developed with the *MS Excel* spreadsheet.

Exemplary time histories of the quantities measured are shown in Figures 17 and 18 and of those calculated - in Figures 19 and 20 for the magnetic bearing of the selected configuration of the control program, at the kinematic excitation of the frequency *40 Hz* and the assigned amplitude, whose value was such as to obtain the dominant share of synchronous components in the time histories under analysis and to obtain the linear range of magnetic response forces.

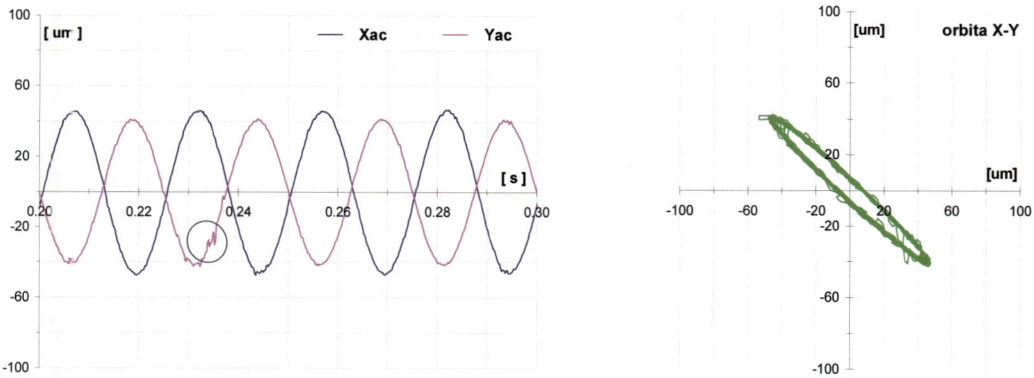

Figure 17. Displacements for both the control axes *X,Y* and the shaft motion trajectory

The occasional disturbances which occur in the recorded time histories of displacements (Figure 17) are amplified by the digital differentiation and the effect of these disturbances is very distinct in the time history of the velocity component V_Y (Figure 19a). This does not affect the calculation accuracy, where the characteristics are approximated with the harmonic time history.

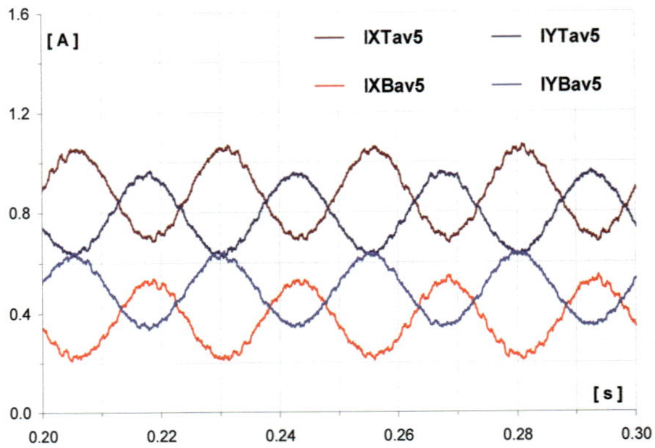

Figure 18. Time histories of the currents in electromagnet windings $I_{XT}, I_{XB}, I_{YT}, I_{YB}$ - averaged

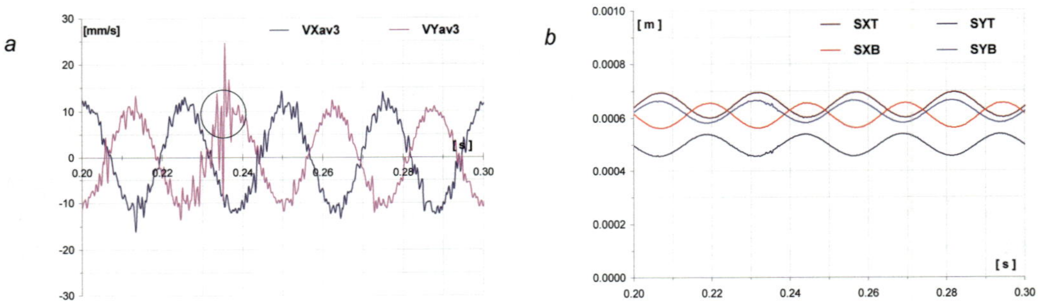

Figure 19. a. – Velocity components for both the control axes V_X, V_Y – averaged
b. – Variable gaps for individual electromagnets $_{XT}, s_{XB}, s_{YT}, s_{YB}$

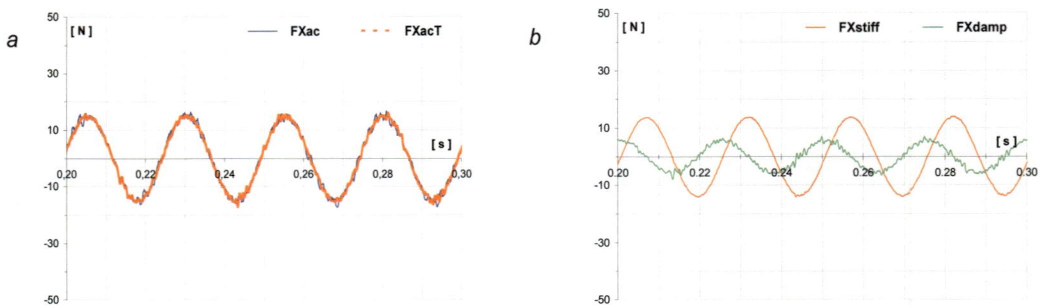

Figure 20. a. – Measured magnetic response component F_{Xac} and the force $F_{XacT} = F_{X\,lin}$ modelled with the identified dynamic coefficients K_{XX} , C_{XX}
b. – Components of the magnetic response related to the stiffness F_{Xstiff} and to the damping F_{Xdamp}

The measurement and calculation cycles were conducted for various excitation frequencies in the range *(10 ÷ 80)Hz*, which allowed one to build functions representing the values of the bearing dynamic coefficients, namely: the stiffness coefficients K_{XX} , K_{YY} and the damping coefficients C_{XX} , C_{YY}, as a function of frequency at the given excitation frequency and the given configuration of the control program (Figures 21).

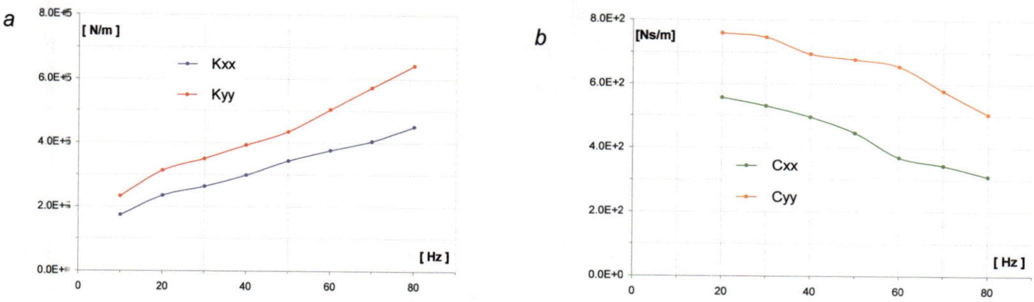

Figure 21. a. – Stiffness K_{XX}, K_{YY} versus frequency b. – Damping C_{XX}, C_{YY} versus frequency

It means that the start-up and shut-down characteristics of the model shaft line under such magnetic bearing operating conditions should be the overcritical characteristics. The experimentally determined dynamic coefficients of the bearing show some anisotropy of the properties for individual axes $K_{XX} \cong K_{YY}$, $C_{XX} \cong C_{YY}$. The analysis of this state showed that for isotropic properties of the energy transmission systems (symmetrical saturation–control current characteristics), this scattering resulted from the scattering of constant values of electromagnets for the given axis. This conclusion was confirmed by the calculations conducted with the simulation model of the magnetic bearing.

The experiment was conducted for various configurations of the program controlling the auxiliary magnetic support used in the model shaft line system. The determined dynamic coefficients were employed in subsequent stages of the investigations in modelling and numerical calculations of start-up and shut-down characteristics of the flexible power-transmission shaft. The generated numerical characteristics were verified experimentally through a comparison to the start-up and shut-down curves recorded for the real model shaft line with supports of identified dynamic properties.

The proposed methodology of measurement of response and dynamic coefficients of the magnetic bearing is a very important tool in designing dynamics and vibration control of machine rotors in which active magnetic bearings are applied. It allows one to find analogues to classical bearing systems and to employ professional calculation codes for evaluation of the effects of modification in the dynamic properties of shaft lines introduced through changes in the configuration of the program controlling its active magnetic supports [4-9].

6. Modification of the dynamics of the rotating system

The last step of this work is to present the experimental results of maintaining a low level of vibrations in the whole operating range of the flexible rotor by the application of an auxiliary active magnetic bearing where the biggest anticipated dynamic deflection of the rotor occurs.

The application of an auxiliary magnetic bearing to modify the dynamics of the rotating system demands the comparison of the classical (without a magnetic bearing) and modified design. In the proposed construction of the test stand, a flexible rotor is supported in ball bearings. An electric engine integrated with the shaft is used as a drive (Figure 22).

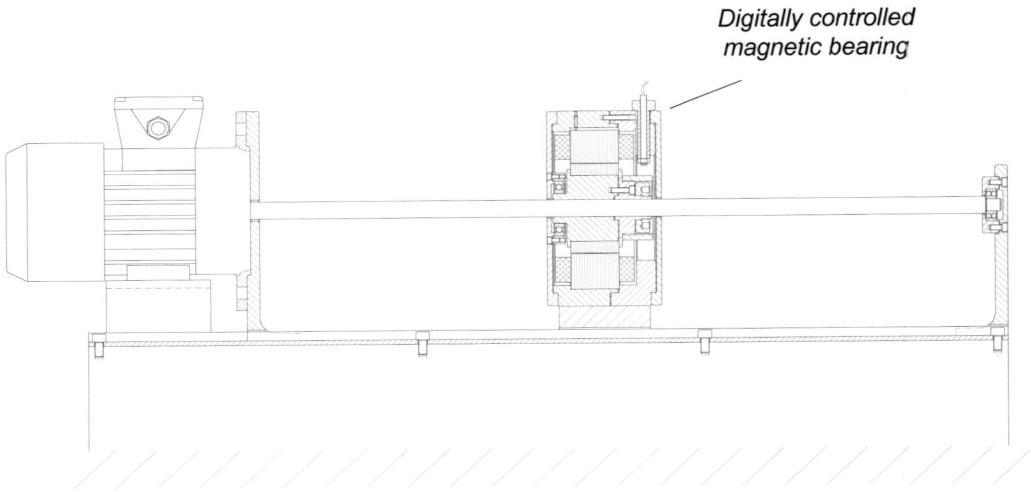

Figure 22. Scheme of the rotating system with an auxiliary active magnetic bearing

The length and the diameter of the shaft is 700 mm and 17 mm, respectively. Each bearing support is connected to the foundation by rigid elements. The mechanical parameters of the designed magnetic bearings are as follows:

- diameter of the journal 89.5 mm
- radial clearance 0.5 mm
- number of electromagnet coil pairs4
- length 45.9 mm
- static load 300 N
- two axes of control x, y with eddy-current transducers.

In this design, the feasible assembly and disassembly of the auxiliary magnetic bearing is assumed. It is the position of the auxiliary magnetic bearing, mounted on the shaft of the test stand, which allows one to model the mode of lateral vibrations. A numerical experiment shows the usefulness of this concept in the case of the test stand rotor. However, the practical application of the presented idea demands serious experimental verification. The aim of this work is to present the results of the experimental investigations of the dynamic response of the system to the synchronous excitation corresponding to the unbalance occurring in the centre of the rotor.

In order to modify the dynamic properties of this structure, an additional journal magnetic bearing should be introduced. The following two cases are considered:

- flexible rotor supported in classical (ball) bearings - the magnetic bearing turned off,
- flexible rotor supported in classical (ball) bearings and an additional magnetic bearing - the magnetic bearing turned on.

When a machine is supported in the whole system of magnetic bearings, the bearings carry the main load that comes from the rotor (lateral, thrust). The role of the magnetic bearing in the system under consideration is to introduce an additional instantaneous point of support for the rotor and to carry a part of the dynamical (lateral) load. Thrust load is always carried by classical (ball) bearings. The proposed additional magnetic bearing does not introduce any additional thrust load. That is why in the proposed analysis, the effect of the active magnetic bearing on thrust force acting on the rotor system isn't under consideration.

The operation of an auxiliary magnetic bearing in the rotating system just before a high level of the amplitude is achieved, causes a qualitative change in dynamic properties of the system. It corresponds to the critical frequency of lateral vibrations and allows for reaching the nominal revolutions of the machine without dangerous effects connected with exceeding the critical frequencies of the flexible rotor.

7. Numerical calculations and verifications

The test stand as presented in Figure 22 was modelled numerically. A professional program *DYROBES* that allows for modelling the dynamics of the shaft line of rotary machines was employed in the numerical calculations. Numerical calculation methods allow one to carry out a complete analysis of machine rotor vibrations. The geometry of the rotating system was modelled with discrete elements. The magnetic bearing journal is situated in the centre of the shaft line (Figure 23).

In Figures 23a and b a shaft line model and calculated modes of critical frequencies of the rotating system for two configurations, i.e. when the magnetic bearing is turned off and when it provides an additional support of the stand, are depicted.

In the configuration under consideration, the dynamic properties of the magnetic bearing are a vital element that decides about a value of the critical frequency of the analysed shaft line. These properties are connected with the assumed parameters of the control system and were identified through the analysis of the bearing simulation model.

The analysis of the rotating system dynamics was performed and the first lateral critical frequency and its respective deflection line of the rotor supported in ball bearings were determined *(38.2Hz)*. The required value of the dynamic stiffness of the magnetic bearing that allows for avoiding the necessity of exceeding the critical value at the start-up of the model shaft line is equal to $K_M > 3E+5\ N/m$.

The second stage of calculations consisted in a determination of the theoretical start-up curve of the modelled shaft line of the test stand. After introducing the linearized coefficients of stiffness and damping of the auxiliary active magnetic bearing, determined on the basis of the model, a response to the rotating synchronous excitation was determined.

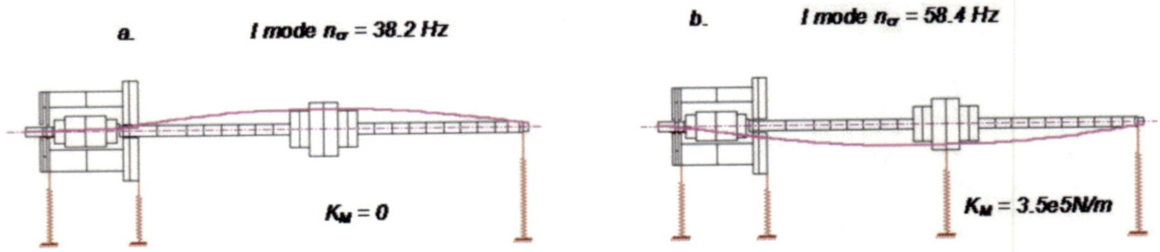

Figure 23. Shaft line model of the test stand and the first critical mode of vibrations a) magnetic bearing turned off, b.) magnetic bearing turned on

The synchronous excitation with a force coming from the residual unbalance was introduced. The values of the vibration amplitude as a function of frequency are the system response. For the first critical frequency, the vibration amplitude reaches the value of *260μm p-p* for the rotating system supported only in ball bearings (Figure 26b). After introducing the auxiliary magnetic bearing in the form of dynamic coefficients of stiffness and damping into the model calculations, the theoretical Bode plot shown in Figure 26 was obtained.

The experimental investigations, like the theoretical ones, were carried out in two stages:

- with a fixed shaft of the test stand,
- during the start-up and shut-down of the model rotating system.

In the first stage, the dynamics of the rotating system was analysed by means of the modal testing method.

For a fixed shaft of the test stand, the tests were carried out for two variants of operation: with the magnetic bearing turned on and turned off.

The system was forced to vibrate by means of the excitation pulse force, whose direction was convergent with the direction of the response measurement (selected control axis). The spectral concentration of this excitation in the frequency range investigated *0÷200 Hz* was practically constant.

The analysis of the time histories (Figure 24a) points out to the fact that in the system with the magnetic bearing turned off, weakly damped oscillations occur as a result of the excitation and they disappear after approximately $t = 3s$.

In the system in which the bearing is activated, these vibrations are strongly damped. In the whole time span observed, slight oscillations of the rotor occur. They result from the principle of bearing operation, which consists in the continuous control of the assumed position of the rotor balance. The time histories were recorded in the time interval $t = 4s$ in order to provide evidence of stable operation of the system after the pulse excitation force had been applied.

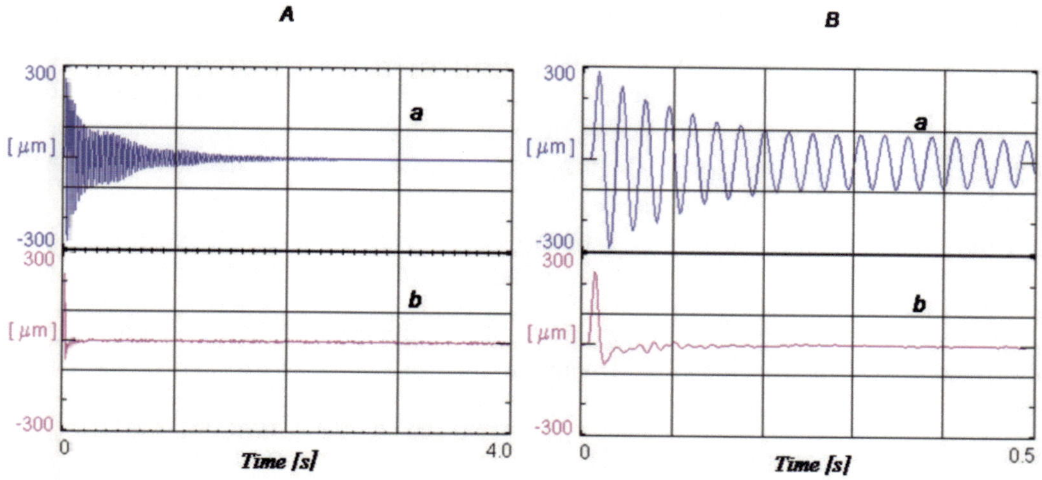

Figure 24. Time history of the response (displacement) recorded along the direction x for the fixed rotor A - recording time $t = 0 \div 4s$, B - recording time $t = 0 \div 0.5s$ a.) auxiliary magnetic bearing turned off, b.) auxiliary magnetic bearing turned on

To show the dynamics of the response of the system with the activated magnetic bearing in Figure 24b, the same time histories of the response are presented for the shorter recording time.

During this time, an effect of the system modification that consists in an activation of the additional magnetic bearing with the programmable assigned dynamic properties is clearly visible. Then, the shaft vibrations disappear after approximately $t = 0.05s$.

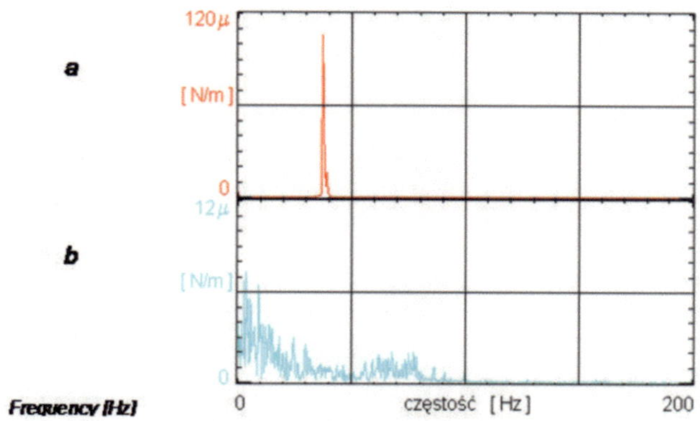

Figure 25. Frequency response for the axis x in the frequency range $n = 0 \div 200Hz$ for the fixed rotor a) auxiliary magnetic bearing turned off, b) auxiliary magnetic bearing turned on

In Figure 25 the transmittances of the rotating system measured on the magnetic bearing journal have been compared.

The analysis of these functions shows that in the system with the bearing turned off, weakly damped vibrations with the frequency of approximately $n = 39$ Hz (the first vibration frequency of the rotor in this configuration - Figure 26a) dominate.

The transmittance of the system with the bearing turned on exhibits strong vibration damping. The applied pulse excitation is not able to generate self-vibrations of the rotor. The recorded frequency response (Figure 25b) is located on the level of noises and equals ≈ *6μm*. The first frequency of self-vibrations of the rotor in this configuration (estimated in the calculations to be equal to approx. *n = 60Hz* - cf. Figure 26b) does not occur in practice in the results of the experimental investigations. The investigations carried out when the rotor did not move have proven a significant influence of the auxiliary magnetic bearing on rotor dynamics.

They have shown possibilities of obtaining an abrupt change in the dynamic properties of the rotating system, which complies with the idea of the safe exceeding of the critical frequency of the machine.

In the second stage of the experimental investigations, the dynamics of the model rotating system with a synchronous excitation with unbalance forces was analysed. The investigations were carried during the start-up and shut-down of the system rotor. Numerical simulations of the test stand rotor response to unbalance were performed. An example of the calculation results shown in Figure 26 indicates that for the nominal rotational frequency *50Hz*, the test stand rotor can be an overcritical or undercritical system, depending on the fact whether the auxiliary magnetic bearing characterised by specified dynamic properties is activated or disactivated.

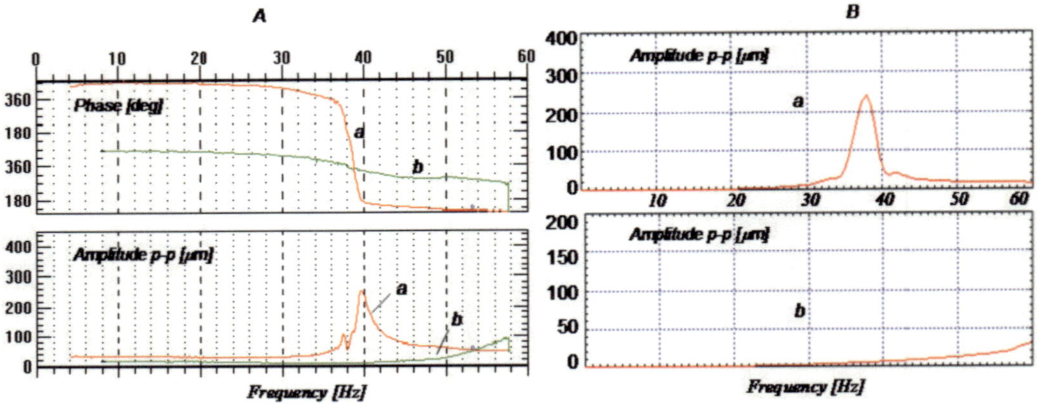

Figure 26. *A* – Experimental Bode plot *B* – Theoretical Bode plot *a* – start-up, the rotating system supported in ball bearings, *b* – shut-down, with the auxiliary magnetic bearing activated

An application of the auxiliary magnetic bearing allowed for a complete elimination of a dangerous increase (for the whole local system operation) in the vibration amplitude when the critical frequency had been exceeded. The optimal moment for activation and disactivation of the additional bearing system has been indicated by a solid line. At this moment, the vibration amplitudes of both the qualitatively different dynamic systems reach similar values. In the presented example, the switching took place above the optimal values of revolutions, which was connected with an abrupt change in the vibration amplitude and phase.

In Figure 27 cascade plots of the journal vibrations in the magnetic bearing during the start-up of the test stand in the so-called "full spectrum" domain are presented. A comparison of the diagrams illustrates the role played by a presence of the auxiliary magnetic bearing from the point of view of shaft vibrations during the start-up (or shut-down) of the model machine.

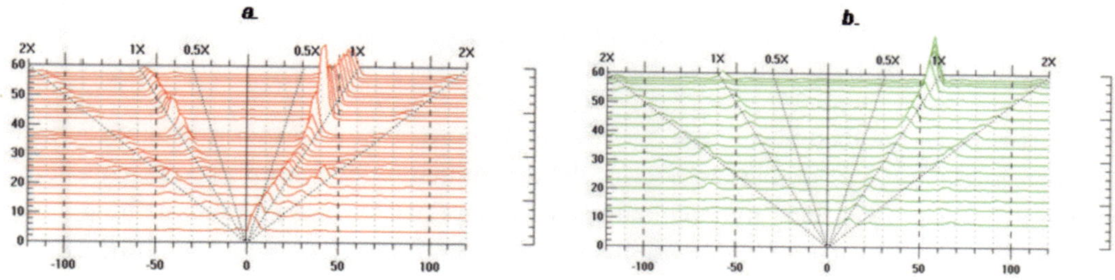

Figure 27. Cascade plots of the journal vibrations in the magnetic bearing during the start-up of the model shaft line a) auxiliary magnetic bearing disactivated, b) auxiliary magnetic bearing activated

The presented idea of an application of the active magnetic bearing as a system that modifies the dynamic properties of the rotating system can turn out to be an interesting alternative to modernisation of a real rotary machine. The majority of overcritical failures of rotors is caused by a local increase in the synchronous or asynchronous vibration amplitude and by exceeding the clearances when the critical frequency or the stability threshold of the rotating system is exceeded. The special character of the active magnetic bearing operation makes it possible to activate it when the shaft rotates, whereas relatively high values of the clearances allow for locating it practically in any place of the machine shaft line. The built test stand enabled practical realisation of the idea of "omitting" the region in which rotor vibrations, dangerous for the machine operation, appear.

8. Conclusions

The progress in the field of active magnetic bearings is based on new operating and diagnostic capabilities of these digitally controlled bearings in comparison to classical solutions. The overall goal of an *AMB* controller is to stabilise the plant and to reach the optimal technical performance. To achieve these goals, *AMB* systems have to be optimised in an overall mechatronics design approach. This leads to a new concept for control systems and actuators [11-14].

In this work, an idea of the simulation model of magnetic levitation systems and its diagnostic capabilities is presented. Some results of the numerical simulations and experimental investigations are discussed.

The conducted investigations allowed one to verify experimentally stiffness and damping of the real rotor-magnetic bearing system by means of a numerically calculated model of the rotor.

The proposed methodology of measurement of response and dynamic coefficients of the magnetic bearing is a very important tool in designing dynamics and vibration control of machine rotors in which active magnetic bearings are applied. It allows one to find analogies to classical bearing systems and to employ professional calculation codes for evaluation of the effects of modification in the dynamic properties of shaft lines introduced through changes in the configuration of the program controlling its active magnetic supports.

A comparison of the theoretical time histories with those obtained experimentally confirms the correctness of the proposed method for the determination of dynamic coefficients of the magnetic bearing. Achieving the nominal speed of the flexible rotor and maintaining a low level of vibrations in the whole operating range of the rotating system (including critical speeds) by using an auxiliary active magnetic bearing is a very interesting idea for the rotating machinery.

The operation lets "omit" the zone of a dangerous increase in the amplitude of rotor vibrations, which is connected with the critical speed of lateral vibrations. The experiment shows the usefulness of this concept in the case of the real rotating system.

The symmetry of characteristics of individual actuator paths of bearings, which has been programmable corrected and experimentally verified, has made it possible to implement an idea of the application of a single controller to control the real journal bearing operation along both axes. A selection of values of controller parameters is based on the investigation of the simulation model of an active magnetic bearing system.

The procedures of numerical representations of the actuator characteristics have allowed for a development of the model whose operation is convergent with the real bearing system. It enables simulation investigations of the dynamics of the mass suspended in the bearing bush under widely variable values of controller parameters and under various disturbances and forces. A reliable theoretical model that allows for analysis of the bearing dynamics under hypothetical, extreme loads reduces the designing time and enables one to minimize errors that can occur at the system prototype start-up.

The experimental characteristic curves of the start-up and shut-down have confirmed possibilities of the programmable modelling of dynamics of the shaft line, and - in prospect - of the designed machine that includes an additional active magnetic bearing.

Author details

Dorota Kozanecka
Institute of Turbomachinery, Lodz University of Technology, Łódź, Poland

9. References

[1] Schweitzer G., Traxler A., Bleuler H., (1993), *Magnetlager,* Springer-Verlag, Berlin (*in German*).

[2] Kozanecka, D., et al., (1997), Application of an Auxiliary Active Magnetic Bearing for Modification of Dynamic Properties of Rotors, *Proc. of the XIV World Congress IMEKO,* June, Finland, Vol. IX, pp. 93-98.

[3] Delprete, C., Genta, G., Repaci, A., (1998), Numerical Simulations of the Dynamic Behaviour of Rotors on Active Magnetic Bearing, *Proceedings of the 7 Int. Symp. on Transport Phenomena and Dynamics of Rotating Machinery,* Honolulu, Hawaii, February, Vol. A, pp. 48-57.

[4] Kozanecka D., (1999), Dynamic Properties of the Rotor Magnetic Suspension System, *Symkom'99* Łodz, Technical University of Łodz, Journal CMP – Turbomachinery, No 115, pp. 217-224.

[5] Kozanecka D., 2001, *Dynamic of the Flexible Rotor with an Additional Active Magnetic Bearing,* Machine Dynamics Problems, 2001, Warsaw, Vol. 25, No. 2, pp. 21-38.

[6] Kozanecka D., Kozanecki Z., 2001, *Modelling the Dynamics of Active Magnetic Bearing Actuators,* Proceedings of the World Multiconference on Systemics, Cybernetics and Informatics, SCI 2001, July 22-25, USA, Vol. IX, Industrial Parts I, pp. 232-235.

[7] Kozanecka D., et al., 2003, *New Concept of the Spin Test System with Active Magnetic Bearing,* Procds. 2nd International Symposium on Stability Control of Rotating Machinery, ISCORMA, Gdansk 4-8 August 2003, Poland, pp. 199-219.

[8] Kozanecka D., et al., (2007), *Application of Active Magnetic Bearings for Identification of the Force Generated in the Labyrinth Seal,* Journal of Theoretical and Applied Mechanics, Polish Society of Theoretical and Applied Mechanics, No.1, Vol. 45, pp. 53-60.

[9] Kozanecka D., et al., (2008), *Experimental Identification of Dynamic Parameters for Active Magnetic Bearings,* Journal of Theoretical and Applied Mechanics, Polish Society of Theoretical and Applied Mechanics, No.124, Vol. 46, pp. 41-50.

[10] Schweitzer G., Maslen E.H., (2009), *Magnetic Bearings. Theory, Design and Application to Rotating Machinery,* Springer-Verlag, Berlin, Heidelberg.

[11] Gizelska M., Kozanecka D., Kozanecki Z., (2009), *Integrated Diagnostics of the Rotating System with an Active Magnetic Bearing,* Solid State Phenomena, Trans Tech Publications Ltd, Switzerland, ISSN 1012-0394, Vol. 147-149 pp 137-142. (Online available since 2009/Jan/06, at http://www.scientific.net)

[12] Łagodziński J., (2009), *Modeling of Magnetic Fields with the Finite Element Method in Machine Diagnostic Systems,* Solid State Phenomena, Trans Tech Publications Ltd, Switzerland, ISSN 1012-0394, Vol. 147–149, pp. 155–160. (Online available since 2009/Jan/06, at http://www.scientific.net)

[13] Kozanecka D., Kozanecki Z., Łagodziński J., (2011), *Active Magnetic Damper in a Power Transmission System,* Communications in Nonlinear Science and Numerical Simulation, Journal Elsevier, No 16, pp. 2273-2278 www.elsevier.com/locate/ cnsns_1567.

[14] Kozanecka D., (2010), *Diagnostics of Rotating Machinery Mechatronic System,* Monographic Series of Publications: Maintenance Problems Library, Scientific Publishing House of Institute for Sustainable Technologies in Radom, ISBN 978-83-7204 966-7, (*in Polish*).

Chapter 8

Control of Magnetic Bearing System

Hwang Hun Jeong, So Nam Yun and Joo Ho Yang

Additional information is available at the end of the chapter

http://dx.doi.org/10.5772/51185

1. Introduction

A bearing fixes a rotating spindle to a specific location and is a mechanical component that supports the load applied to the axis and its dead load. Therefore, it is inevitable for mechanical contact between the axis and the bearing to occur, causing friction, abrasion, heat, noise, and user environment contamination from lubrication. Magnetic bearings are mechanical components that use the attractive or repulsive force of electromagnets to support the mechanical axis is a non-contact state. The use of such components significantly reduced the disadvantages that accompany the use of general mechanical bearings such as friction, abrasion, heat, noise, and user environment contamination from lubrication. Moreover, magnetic bearings can support the mechanical axis in special environments(vacuum, high temperature, low temperature, zero gravity) and have the advantage of being able to adjust the damping coefficient and spring constant of the system that supports the axis according to the control objective.

Magnetic levitation can be categorized into the following systems depending on the form of force that supports the levitated object: the system that uses magnetic attraction, magnetic repulsion, induction levitation, and superconducting Meissner Effect. Magnetic levitation that utilizes attractive force has a closed magnetic circuit so efficiency is high and 1-axis control is possible due to the stability in the attraction and perpendicular directions. However, it has been reported that the uncontrolled directions have poor stability due to the nonlinearity of the attraction. Magnetic levitation that uses the repulsive power has stable characteristics with respect to the longitudinal direction that the repulsive force is applied to, but the transverse direction has unstable characteristics. However the electromagnet is arranged, all the axes cannot be stabilized. Magnetic levitation that uses induction levitation is able to perform stable levitation without special control as Fleming force caused by the relative velocities between the electromagnet and the conductor supports the levitation. However, without a velocity over a certain level, levitation cannot be supported where overall efficiency is low due to Eddy current loss and brake loss. Magnetic levitation that uses superconducting Meissner Effect takes advantage of the repulsion with permanent magnets caused by the strong diamagnetism from the superconductor. Like that of the induction levitation, stable levitation is possible without any control. However, the operational temperature of the levitation system using superconductor is very low: 4.2K(liquid Helium), 77K(liquid Nitrogen). Generally, the magnetic levitation applied to the magnetic bearing is the method using attraction and repulsion. Magnetic bearing systems discussed here refers to a system that utilizes the attraction.

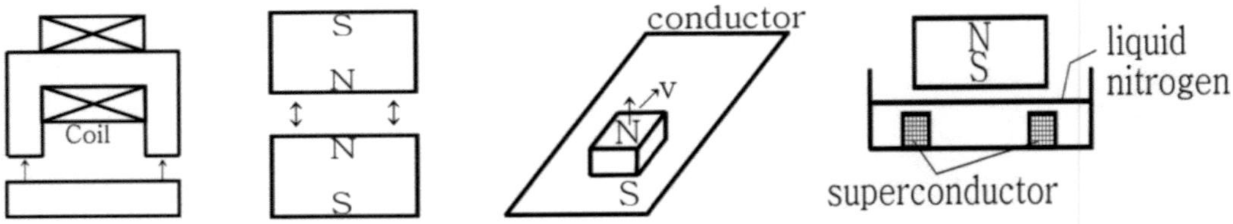

Figure 1. Levitation system according to the magnetic levitation method

In this chapter, the method to designing a magnetic bearing system, to obtaining a mathematical model, and understanding the preparations necessary for control will be discussed. For this, the calculation of attraction using the Probable Flux Paths Method, selection of circuitry about the amplifier to operate the electromagnet, and method to identify the magnetic bearing system that includes a PID controller are discussed.

2. Magnetic bearing system design

Fig. 2 shows a schematic diagram of the magnetic bearing system to be designed in this chapter. The levitated object is supported by the attraction of the electromagnet and the attraction of the electromagnet is controlled by the current in the coil. In order to design and control such a magnetic bearing system, the amplifier to operate the electromagnet that composes the magnetic bearing system and hardware to control the whole system need to be designed first. Next, the designed magnetic bearing system is modeled mathematically, then the parameter values difficult to measure through the mathematical model are determined through experimentation. Finally, an adequate controller is designed and applied to the identified magnetic bearing system. In this chapter, the detailed control laws for magnetic bearing control are excluded and the implementation of the magnetic bearing system before applying various controllers is the main objective.

2.1. Magnetic bearing system composition

A magnetic bearing system like that of Fig. 2 is composed of the object to be levitated, core, electromagnet including the coil, amplifier to operate the electromagnet, displacement measurement system to measure the distance between the levitating object and the electromagnet, control law to calculate the control signal from the feedback signal, and control system that includes the hardware to realize the control law.

Fig. 3 is the assembly of the magnetic bearing system to make. In the assembly, the levitated object will be supported by magnetic bearing at X and Y axis direction. But thrust direction has only the mechanical backup bearing. And Fig. 4 is the levitated object that is 1.4kg.

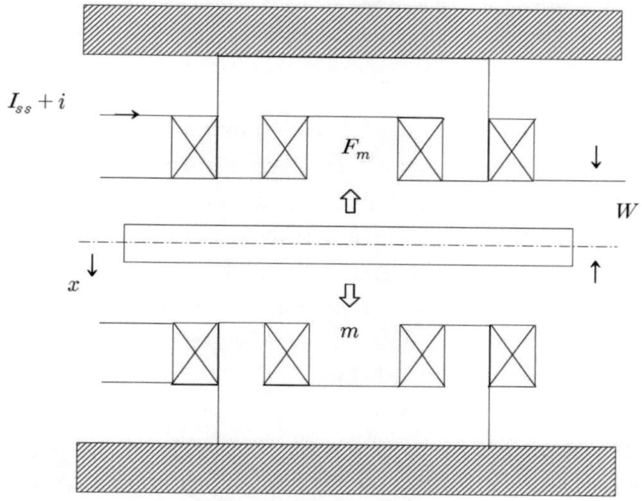

Figure 2. Schematic of the magnetic bearing system

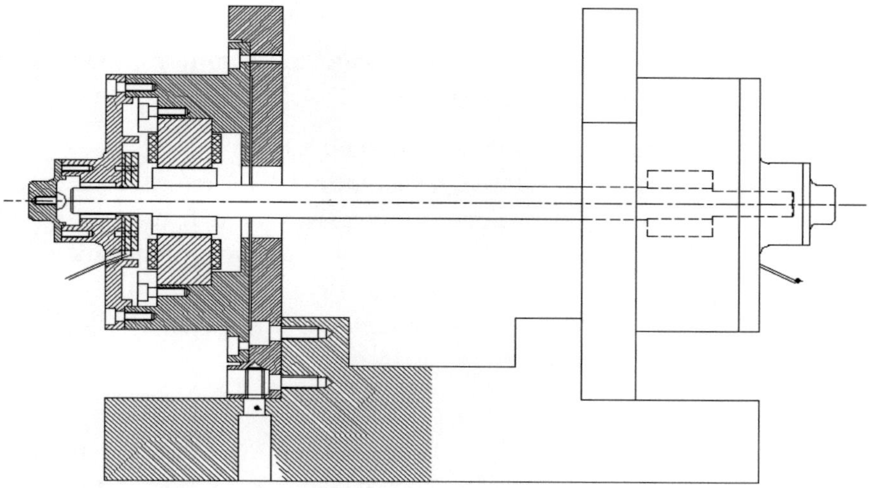

Figure 3. Assembly of the magnetic bearing system

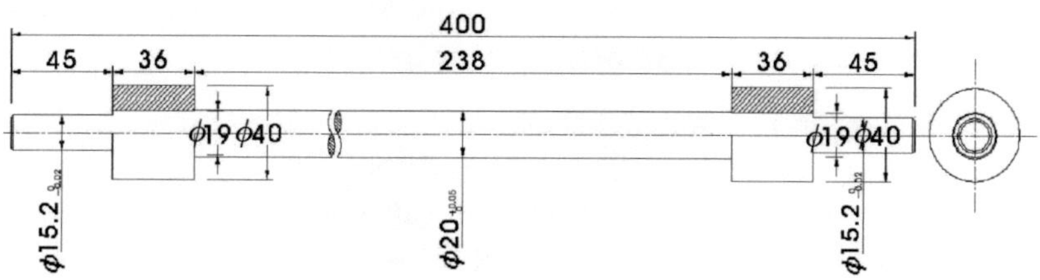

Figure 4. Levitated object

Since a magnetic bearing system like Fig. 3 has a symmetric form vertically and horizontally, the levitating object can be simply assumed as a point mass in the perspective that the object is levitated. Therefore, in this section, the elements of the magnetic bearing system excluding the levitating object are designed.

2.1.1. Probable flux paths method

To support the levitating object through the electromagnet's attractive force, the attraction relationship between the current in the electromagnet coil and the levitating object needs to be defined clearly. The Probable Flux Paths Method assumes that the magnetic permeability of the magnetic substance that forms the magnetic path is linear to calculate the permeance of the magnetic substance that the magnetic path passes through, followed by the calculations of the magnetomotive force, magnetic flux, magnetic flux density, and attractive force. As the permeability of materials disregarding permanent magnets are generally nonlinear, the error between the Probable Flux Paths Method calculation results and that of actual experimentation measurements is large and calculations by applying the Probable Flux Paths Method for magnetic substances with complicated magnetic paths is known to be difficult. However, since the vertically and horizontally symmetric magnetic bearing system magnetic path is of a simple form, the electromagnet is designed by applying the Probable Flux Paths Method early in the design process. For more precise designing, the use of FEM software such as Maxwell is desirable.

Generally, the following assumptions have to be satisfied when using the Probable Flux Paths Method to analyze the magnetic circuit.

a. The relationship between magnetic flux and current is linear.
b. The average magnetic flux passes through the centroid of the cross section.
c. When the cross section that the magnetic flux passes through changes, the parts are calculated by dividing them into different parts and setting as combinations of parallel or series.
d. When the cross section of a part changes rapidly, the magnetic flux passes through in a smooth circular arc(quadratic curve).

2.1.2. Electromagnet design

In order to design the electromagnet using the Probable Flux Paths Method, first, the attractive force derived from the magnetic circuit caused by the electromagnet needs to withstand the weight of the mass. Here, the following steps in design are taken so that sufficient attractive force from the electromagnet is produced for control.

a. The mass of the levitating object is determined.
b. The material of the core and levitating object is determined.
c. The attractive force of the electromagnet is calculated with values determined by assumptions regarding the current, magnetic circuit, and length of the coil and number of windings during normal conditions.

d. The material of the core and levitating object, coil and number of windings, and current at normal state is adjusted until the comparison of the calculated attractive force value and the mass of the levitating object gives a satisfactory attractive force.
e. The electromagnet's maximum attractive force and coil's maximum current according to the core and saturation flux density of the levitating object material are calculated.
f. The material of the core and levitating object, coil and number of windings, and current at normal state is adjusted until the calculated maximum attractive force of the electromagnet is sufficient.

Since the attractive force calculated through the Probable Flux Paths Method has a large error with the actual experimentation values, in order to manufacture magnetic bearings based on this design, it is desirable to design with a safety factor of greater than 3.

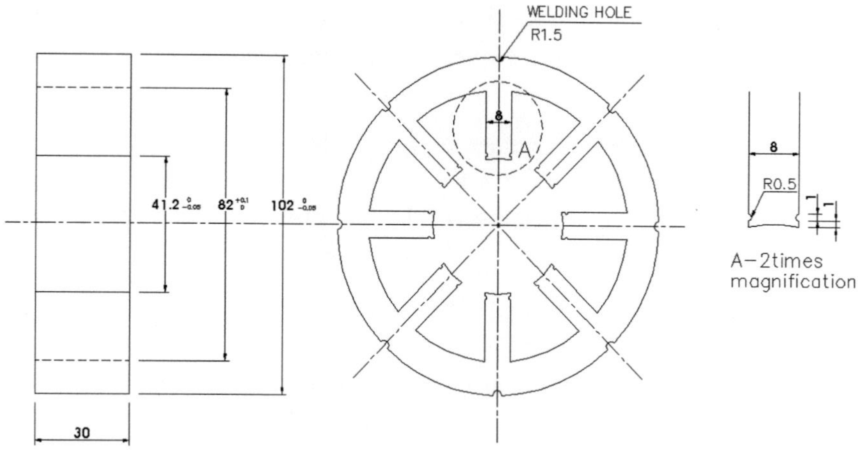

Figure 5. Electromagnet core

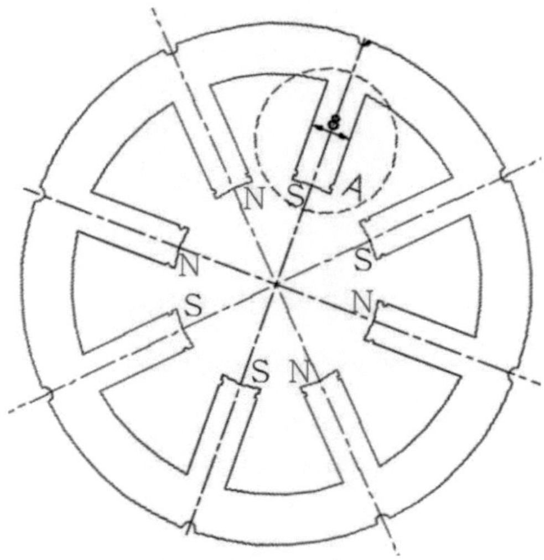

Figure 6. Electromagnet magnetic circuit formation

Next, the number of poles and the angle of the core have to be kept in mind. Fig. 5 shows the core to be used as the electromagnet in the magnetic bearing system. The magnetic circuit caused by the electromagnet that supports one axis has to be designed so that it does not interfere with the magnetic circuit caused by the electromagnetic that supports the other axis. The direction of the magnetic circuit caused by the electromagnet is determined by the direction of the coil wound about the core. The winding has to have polarity as shown in Fig. 6 so that interference of the magnetic circuit does not occur. Also, for the convenience of control, the resultant force of the attraction caused by the magnetic path needs to be parallel or perpendicular to the supporting axis.

2.1.3. Displacement measurement system design

When the current flowing in the coil is constant, the attractive force is proportional to the square of the distance to the levitating object. Therefore, in order to implement a magnetic bearing control system, the distance between the electromagnet and the levitating object has to undergo feedback.

Sensors that measure gaps in noncontact state include using the change in capacitance, change in Eddy Current, and using laser or ultrasound. A displacement measurement sensor has to be selected with consideration of the sampling time, range and area of the gap to be measured, and economic feasibility of the overall system.

2.1.4. Control system design

The objective of a magnetic bearing control system is producing a control signal from the error signal between the reference input and the gap(between the electromagnet and the levitating object). And from this control signal, control the current of the electromagnet coil to reach the reference input in a stable manner. The control period of a control system is the time consumed in performing one-iteration of computations given by the control system. In order to implement a magnetic bearing system of high speed rotations, the control period has to be as small as possible. To implement such a system, the system is designed taking into consideration all the speeds of the MPU which will operate processing the response speed of the displacement sensor, A/D converter speed, D/A converter speed, and discretize the control laws.

2.2. Magnetic bearing system mathematical modeling

2.2.1. Relation between electromagnet and levitating object

First, the mathematical model for the case of the above existing electromagnet supporting the levitating object is derived. The equation of translational motion for the levitating object is as shown in Equation (1).

$$m\frac{dv}{dt} = mg - F_m \tag{1}$$

Here, m is the levitating object mass, is g the gravitational acceleration and F_m is the attractive force of the electromagnet.

Generally, the electromagnet's attractive force is as shown in Equation (2).

$$F_m = \frac{B^2}{2\mu_0} A_m \tag{2}$$

Here, B is the magnetic flux density, μ_0 is the relative permeability of the vacuum and A_m is the opposing area of the electromagnet. Thus, the attractive force of the electromagnet is determined by the opposing area of the electromagnet and the magnetic flux density.

The magnetic circuit and magnetic flux by the electromagnet is as shown in Equation (3).

$$\Phi = \frac{F_{mf}}{R_m} = \frac{NI_m}{\frac{l_m}{\mu S}} = \frac{NI_m}{\frac{l_m}{\mu_0 \mu_s S} + 2\frac{x_0}{\mu_0 S}} = \frac{\mu_0 SNI_m}{\frac{l_m}{\mu_s} + 2x_0} \tag{3}$$

Here, F_{mf} is the magnetomotive force, N is the number of coil windings, R_m is the magnetic resistance, l_m is the length of the magnetic path, μ is the relative permeability, x_0 is the gap between the electromagnet and the levitating object, μ_s is the relative permeability of the metal pin and S is the cross sectional area of the metal pin.

The magnetic flux density of the electromagnet is the magnetic flux per unit area and is as shown in Equation (4).

$$B = \frac{\Phi}{S} = \frac{\mu_0 NI_m}{\frac{l_m}{\mu_s} + 2x_0} \tag{4}$$

Therefore, the attractive force of the electromagnet is as shown in Equation (5).

$$F_m = \frac{\mu_0 N^2 I_m^2 S}{\left(\frac{l_m}{\mu_s} + 2x_0\right)^2} \tag{5}$$

If the electromagnetic coil current in neutral state is assumed as I_{ss} and the varying control current signal from the neutral state is assumed as i, Equations (5) and (6) can be put together and Equation (7) is satisfied.

$$F_m = k\left(\frac{I_{ss}+i}{X_0+W+x}\right)^2 \tag{6}$$

$$mg - k\left(\frac{I_{ss}}{X_0+W}\right)^2 = 0 \tag{7}$$

Here,

$$X_0 = \frac{I_{ss}}{2\mu_s} \text{ and } k = \frac{N^2 \mu_0 S}{4}.$$

To apply linear control theory, the control subject also has to be a linear control system. However, Equation (1) includes a nonlinear term, thus, linear control theory cannot be applied. So, the nonlinear term of Equation (1) is linearized through Taylor series.

Assuming that the values after the second-order Taylor series terms are sufficiently small compared to the first-order term, the Taylor series in the parallel point($i = 0,\ x = 0$) with regard to the nonlinear term $k\left(\frac{I_{ss}+i}{X_0+W+x}\right)^2$ is as shown in Equation (8).

$$k\left(\frac{I_{ss}+i}{X_0+W+x}\right)^2 = k\left(\frac{I_{ss}+i}{X_0+W+x}\right)^2\Bigg|_{i=0,x=0} + \frac{\partial}{\partial i}k\left(\frac{I_{ss}+i}{X_0+W+x}\right)^2\Bigg|_{i=0,x=0}(i-0) + \frac{\partial}{\partial x}k\left(\frac{I_{ss}+i}{X_0+W+x}\right)^2\Bigg|_{i=0,x=0}(x-0)$$

$$k\left(\frac{I_{ss}+i}{X_0+W+x}\right)^2 = k\left(\frac{I_{ss}}{X_0+W}\right)^2 + \frac{2kI_{ss}}{(X_0+W)^2}i - \frac{2kI_{ss}^2}{(X_0+W)^3}x \tag{8}$$

Equation (8) is substituted into Equation (1) to obtain Equation (9) the equation of translational motion of the levitating object.

$$m\ddot{x} = mg - \left\{k\left(\frac{I_{ss}}{X_0+W}\right)^2 + \frac{2kI_{ss}}{(X_0+W)^2}i - \frac{2kI_{ss}^2}{(X_0+W)^3}x\right\} \tag{9}$$

Taking into consideration the conditions of Equation (7), Equations (9) can be rearranged into the following Equation (10).

$$m\ddot{x} = \frac{2kI_{ss}^2}{(X_0+W)^3}x - \frac{2kI_{ss}}{(X_0+W)^2}i \tag{10}$$

The relationship between the current flowing in the coil and the drop of electric pressure in the coil of the upper side electromagnetic in the magnetic bearing is shown as Equation (11).

$$\frac{d}{dt}\left(L(I_{ss}+i)\right) + R(I_{ss}+i) = E_{ss} + e \tag{11}$$

Here, E_{ss} is the voltage that appears due to the current flowing in the electromagnet coil in neutral state, e is the voltage that appears due to the control current flowing in the electromagnet coil, and Equation (12) is satisfied.

$$RI_{ss} = E_{ss} \tag{12}$$

Additionally, the inductance of the coil is proportional to the number of coil windings and the magnetic flux as shown in Equation (13).

$$L = \frac{N\Phi}{I_m} \tag{13}$$

When assuming that there is no leakage magnetic flux in the magnetic path caused by the coil, the coil inductance is as shown in Equation (14).

$$L = \frac{Q}{X_0+W+x} \tag{14}$$

Here, W is the gap between the electromagnet at neutral state and the levitating object and x is the gap between the electromagnet and levitating object varying due to the control input. Moreover, $X_0 = \frac{I_{ss}}{2\mu_s}$ and $Q = \frac{N^2\mu_0 S}{2}$.

If L_0 is the leakage magnetic flux, the coil inductance is as shown in Equation (15).

$$L = \frac{Q}{X_0+W+x} + L_0 \tag{15}$$

Therefore, the term $\frac{d}{dt}\left(L(I_{ss}+i)\right)$ in Equation (11) can be solved like that of Equation (16), and Equation (11) is the same as Equation (17).

$$\frac{d}{dt}\left(L(I_{ss}+i)\right) = L\frac{d}{dt}(I_{ss}+i) + (I_{ss}+i)\frac{dL}{dt}$$

$$\frac{d}{dt}\left(L(I_{ss}+i)\right) = L\frac{di}{dt} + (I_{ss}+i)\frac{\partial L}{\partial x}\frac{dx}{dt}$$

$$\frac{d}{dt}\left(L(I_{ss}+i)\right) = L\frac{di}{dt} + (I_{ss}+i)\frac{-Q}{(X_0+W+x)^2}\frac{dx}{dt} \tag{16}$$

$$L\frac{di}{dt} - \frac{Q}{(X_0+W+x)^2}\dot{x}(I_{ss}+i) + R(I_{ss}+i) = E_{ss} + e \tag{17}$$

Equations (17) and (18) can be obtained from the conditions of Equation (12).

$$L\frac{di}{dt} - \frac{Q}{(X_0+W+x)^2}\dot{x}(I_{ss}+i) + Ri = e \tag{18}$$

If the levitating object is assumed close to neutral state($i = 0$, $x = 0$) and I_{ss} is sufficiently larger than i, Equation (18) is the same as Equation (19).

$$L\frac{di}{dt} - \frac{Q}{(X_0+W)^2}\dot{x}I_{ss} + Ri = e \tag{19}$$

2.2.2. *Relationship of linear amplifier*

Fig. 7 shows the electric circuit of the amplifier to operate the electromagnet coil.

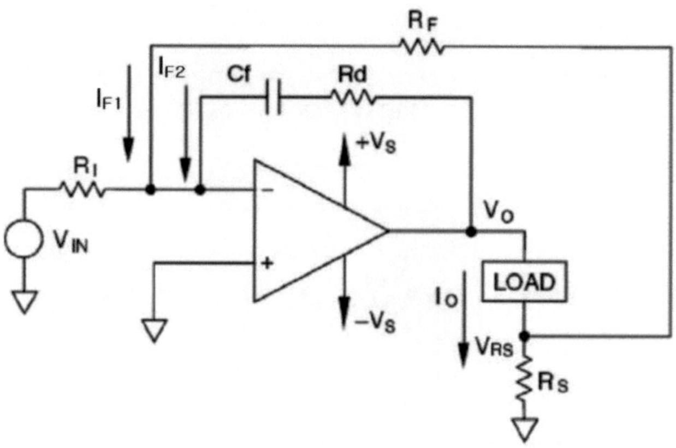

Figure 7. Current amplifier circuit

Assuming the current amplifier as an ideal amplifier, the transfer function from the amplifier control input V_{IN} to the load current I_o is found to be as shown in Equation (20).

$$-\frac{R_I}{R_F}R_S I_o - \frac{R_I}{Z_I}(Z_L + R_S)I_o = V_{IN}$$

$$-\left(\frac{R_I R_S}{R_F} + \frac{R_I(Z_L+R_S)}{Z_I}\right) I_o = V_{IN}$$

$$-\left(\frac{R_I R_S Z_I + R_F R_I (Z_L+R_S)}{R_F Z_I}\right) I_o = V_{IN}$$

$$\frac{I_o}{V_{IN}} = -\frac{R_F Z_I}{R_I R_S Z_I + R_F R_I (Z_L+R_S)} \tag{20}$$

In this circuit, when the impedance Z_I and load impedance Z_L undergoes Laplace Transformation for the transient characteristic improvement of the amplifier, Equations (21) and (22) are obtained.

$$Z_I = \frac{R_d C_f s + 1}{C_f s} \tag{21}$$

$$Z_L = Ls + R \tag{22}$$

Substituting Equations (21) and (22) into Equation (20) and rearranging the equation, the transfer function from the amplifier control input V_{IN} to the load current I_o can be found like Equation (23).

$$\frac{I_o}{V_{IN}} = -\frac{R_F \frac{R_d C_f s+1}{C_f s}}{R_I R_S \frac{R_d C_f s+1}{C_f s} + R_F R_I (Ls+R+R_S)}$$

$$\frac{I_o}{V_{IN}} = -\frac{R_F (R_d C_f s+1)}{R_I R_S (R_d C_f s+1) + R_F R_I (Ls+R+R_S) C_f s}$$

$$\frac{I_o}{V_{IN}} = -\frac{R_F R_d C_f s + R_F}{R_F R_I L C_f s^2 + (R_I R_S R_d + R_I R_F (R+R_S)) C_f s + R_I R_S} \tag{23}$$

2.2.3. *Block diagram and transfer function of the overall system*

The block diagram of the magnetic bearing system using an upper electromagnet from the relationships of Equations (10), (19), and (23) is shown in Fig. 8.

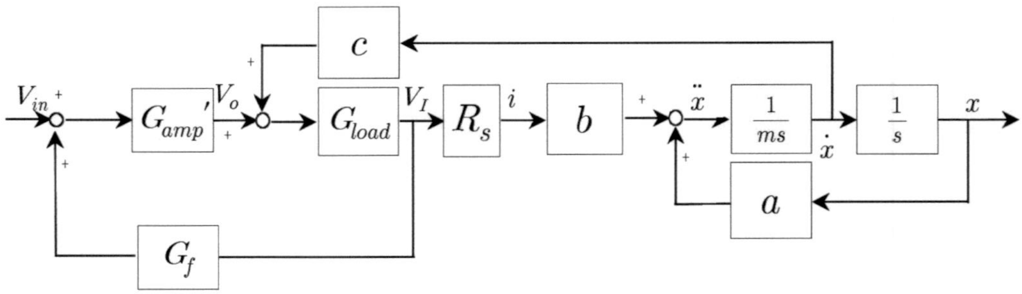

Figure 8. Block diagram of the magnetic bearing system

Here, the transfer function G_{MB} from the current amplifier circuit input voltage V_{IN} to the air gap x between the magnetic bearing and the levitating object is as shown by Equation (24).

$$G_{MB} = \frac{b_{mb1}s+b_{mb0}}{a_{mb4}s^4+a_{mb3}s^3+a_{mb2}s^2+a_{mb1}s+a_{mb0}} \tag{24}$$

Here,

$$
\begin{aligned}
a_{mb4} &= R_I R_F C_f L m \\
a_{mb3} &= R_I R_F C_f m(R + R_s) + R_I R_d C_f m \\
a_{mb2} &= R_I m - R_I R_F C_f(bcR_s + aL) \\
a_{mb1} &= -a(R_I R_F C_f(R + R_s) + R_I R_d C_f) \\
a_{mb0} &= -aR_I \\
b_{mb1} &= -bR_F R_d C_f \\
b_{mb0} &= -bR_F \\
a &= \frac{2kI_{ss}^2}{X_1^3} \\
b &= -\frac{2kI_{ss}}{X_1^2} \\
c &= \frac{2kI_{ss}}{X_1^2} \\
X_1 &= X_0 + W + x \\
X_0 &= \frac{l_m}{2\mu_s}.
\end{aligned}
$$

3. Magnetic bearing system control

To control the mathematically modeled magnetic bearing system, a process to design the driver to operate the electromagnet and a process to identify unknown parameters are necessary. Here, the method to determine the peripheral device values of the linear amplifier circuit that has the desired output by applying a genetic algorithm and the method to identify the magnetic bearing system using a PID controller to stabilize the genetic algorithm and system are discussed.

3.1. Genetic algorithm

In the method to design a linear amplifier with an output sought by the designer or a method to identify the system parameters through random experimentation data, there are methods available using a frequency response method and applying genetic algorithm. Here, the method to apply genetic algorithm and selecting the desired value is explored.

Genetic algorithm is an algorithm that imitates genetics and natural evolution to optimize the objective function and find the solution set with a structure as shown in Fig. 9.

Genetic algorithm initially generates an initial group to solve the optimization problem defined mathematically. Through the difference with the objective function, the degree of agreement of the chromosomes in the generated initial group is calculated and the result of the calculation becomes the basis for dividing the chromosomes into dominant and recessive chromosomes. Through the reproduction operation based on the degree of agreement of the initial group, they become the source of breeding and through crossover operation, a temporary population is generated. Mutation operation on the generated temporary population leads to the generation of the next generation population. The process of generating the next generation population after going through the aforementioned series of processes is described as one generation and the method to finding an optimized solution to an objective function through operations in a specific generation is defined as an algorithm.

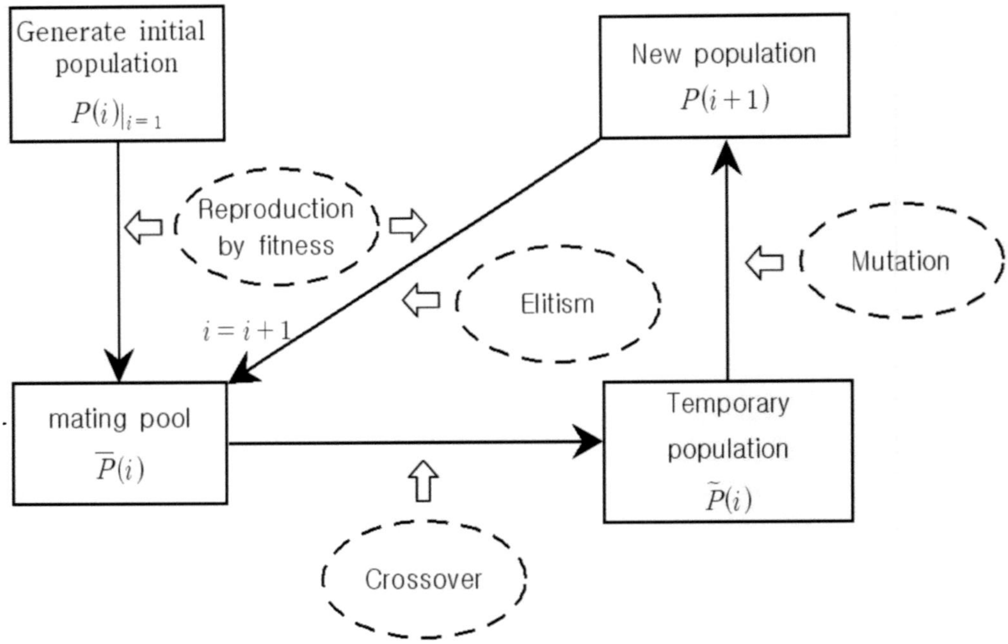

Figure 9. Genetic algorithm

Genetic algorithms can be categorized into BCGA(Binary Coded Genetic Algorithm), SGA(Signal Genetic Algorithm), and RCGA(Real Coded Genetic Algorithm) depending on the expression of the chromosome. Generally, RCGA is used for optimization problems regarding continuous search domain variable with constraints. This is because if the chromosome is expressed by real code, genes that match perfectly with the variable in question could be used and the degree of precision of the calculation is only dependent on the calculation ability of the computer regardless of the length of the gene.

3.2. Amplifier peripheral circuit design

As can be seen in Fig. 7, the linear amplifier circuit is composed of resistance R_I which determines the amplification ratio of the current amplifier and the amplifier output current, resistance R_F, resistance R_d which determines the dynamic characteristics of the linear amplifier circuit, condenser C_f, R_s which limits the linear amplifier circuit current, and the load(here, coil). When defining the form of output desired by the designer using time response characteristics, the amplifier peripheral device values can be determined in the following manner utilizing genetic algorithm.

First, the value of the part to solve is defined. Since the amplification ratio A of the current amplifier, current limiting resistance R_s, load inductance L, and resistance R are unknown, the variables of the genetic algorithm to find are limited to the resistance R_I that determines the amplification ratio of the amplifier output current, resistance R_F, resistance R_d which determines the dynamic characteristics of the linear amplifier current, and condenser C_f. At this point, if the amplification ratio of the amplifier output current is given, one less genetic algorithm variable needs to be found as the resistances R_I and R_F have a proportional relationship.

Next, the searching range of the parameters to be identified is limited according to the characteristics of each device. As the resistance R_I which determines the amplification ratio of the amplifier output current is a signal resistance, it is desirable to have a high resistance value. Therefore, in the case of resistance R_I which determines the amplification ratio of the amplifier output current, it has to be sought in the kΩ range. In contrast, resistance R_d which determines the dynamic characteristics of the linear amplifier circuit has to be sought in a wide range. For condenser C_f, which determines the dynamic characteristics of the linear amplifier circuit, a value in the nF to μF range is ideal when considering the dynamic characteristics of the current amplifier.

After that, the objective function is determined to implement the genetic algorithm. The objective in this program is the design of a linear current amplifier that has a current output in the form that the designer seeks. Therefore, it has the form shown in Equation (23) and the response of the system that satisfies the time response characteristics defined by the designer is defined as shown in Equation (25).

$$G_r(s) = -\frac{e_1 s + e_0}{s^2 + d_1 s + d_0} \tag{25}$$

At this point, the randomly given d_0, d_1, e_0 and e_1 are the coefficients of the system G_r that satisfies the time response characteristics. The objective function to implement the genetic algorithm is defined as shown in Equation (26).

$$F_{obj} = \int e(t)^2 dt \tag{26}$$

Here, $e(t) = g_r(t) - g_{amp}(t)$, $g_r(t)$ is the step response of the system defined by the designer and $g_{amp}(t)$ is the step response of the current amplifier transfer function.

Finally, the parameters to operate the genetic algorithm, such as the size of the entity group, the maximum chromosome length, maximum number of generations, crossbreeding probability, and mutation probability, are defined. Here, in order to improve the performance of the implemented genetic algorithm, configuration for methods such as the penalty strategy, elite strategy, and scale fitting method is necessary.

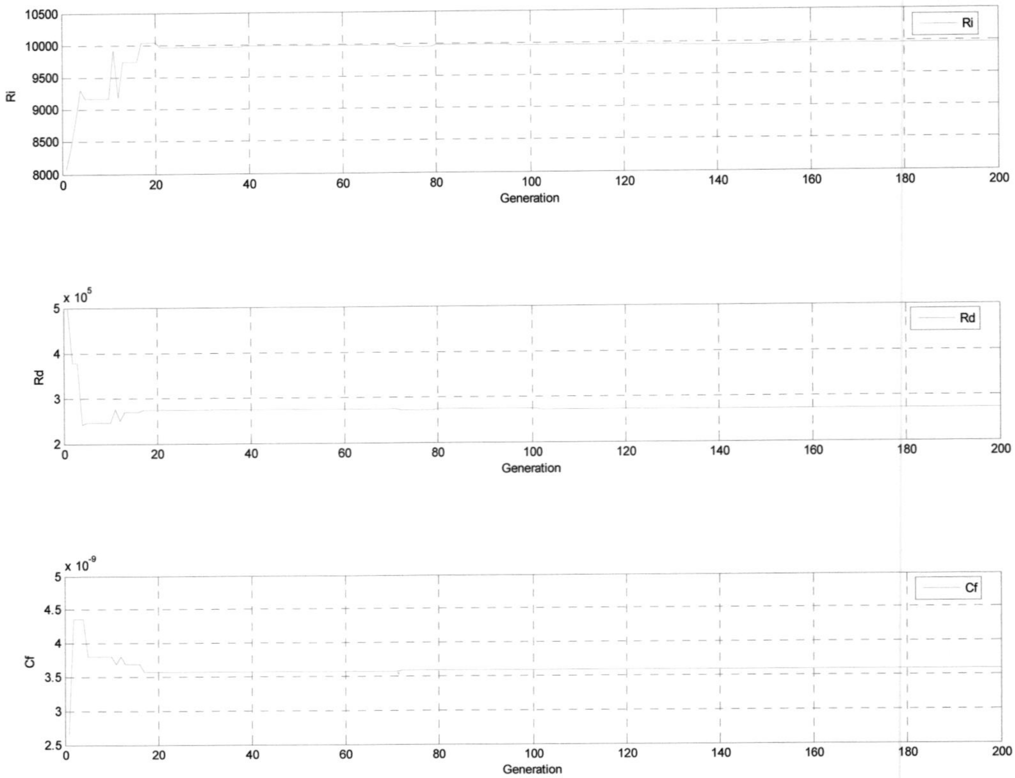

Figure 10. Trends of the optimum parameters of each generation

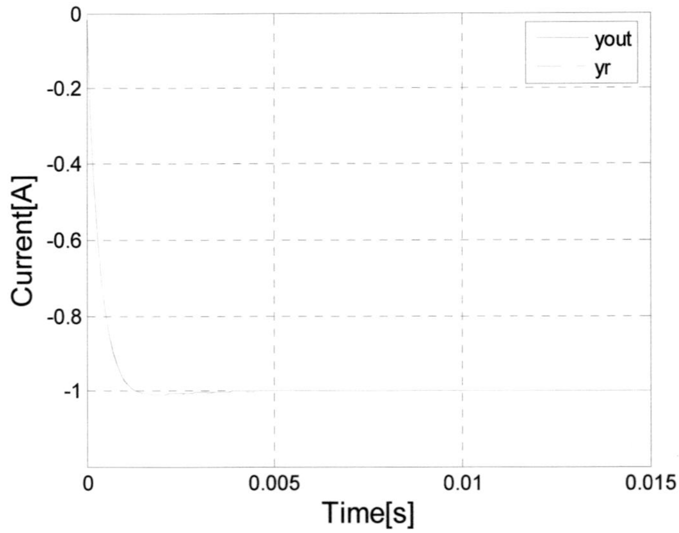

Figure 11. Step response comparison between the system obtained from the RCGA results and the system defined by the designer

Figure 12. Amplifier applied with the designed parameter value

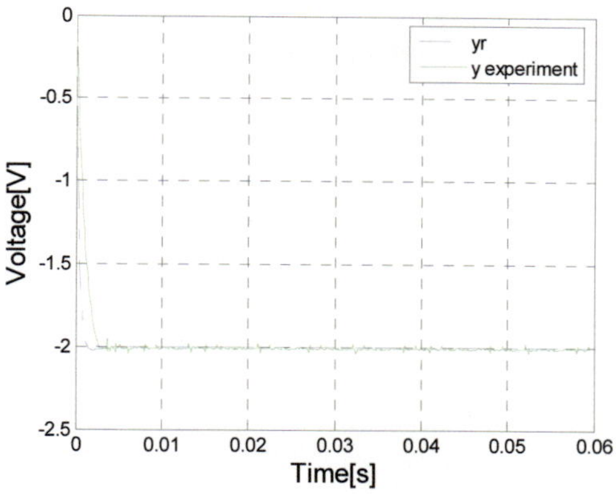

Figure 13. Step response comparison between the system that obtained from the RCGA results and the manufactured amplifier

Appendix 1 is a program that estimates the amplifier peripheral circuit part values in accordance with the response of the system defined arbitrarily, and the result is shown in Fig. 10, Fig. 9, Fig. 11 and Fig. 12 shows the step response measurement graphs of the system that was produced by designing and manufacturing the amplifier circuit using RCGA like that of Appendix 1.

3.3. Magnetic bearing system identification

In order to design the controller for the designed magnetic bearing system, there needs to be a process to estimate the parameter values that exist in the given system and is difficult to measure. In case a magnetic bearing system model with the same response as that of experimentation results can be found, the designer can design the desired controller without performing experimentation. Such a process is called identification.

In order to identify the magnetic bearing system, experimentation data of the manufactured system is necessary. However, since the magnetic bearing system is an unstable system, a controller to stabilize the system to obtain experimental data from the experiment device is necessary.

3.3.1. Implementation of the control signal

To control the magnetic bearing system, there is a need to stabilize the whole system by controlling the current flowing in the current. To do this, PID controller is introduced. In order to control the current of the coil using a PID controller, the voltage going into the current amplifier has to be controlled, and to implement such a voltage signal, a PID controller that is supported by Labview or Matlab has to be used or one that is discretized to match the sample time has to be used. Here, how a discretized PID controller is implemented is described.

Generally, a given PID controller can be defined as shown in Equation (27).

$$\frac{u(s)}{e(s)} = K_p + \frac{K_I}{s} + K_D s \tag{27}$$

Converting Equation (27) to z results in Equation (28).

$$\frac{u(z)}{e(z)} = K_p + \frac{K_I T(z+1)}{2(z-1)} + \frac{K_D(z-1)}{Tz} \tag{28}$$

Reducing Equation (28) and reorganizing u(z) about gives Equation (29).

$$u(z) = \frac{2K_p Tz(z-1) + K_I T^2 z(z-1) + 2K_D(z-1)^2}{2Tz(z-1)}$$

$$(z^2 - z)u(z) = \left(\left(K_P + \frac{T}{2}K_I + \frac{K_D}{T}\right)z^2 + \left(\frac{T}{2}K_I - K_P - \frac{2K_D}{T}\right)z + \frac{K_D}{T}\right)e(z) \tag{29}$$

Taking into consideration that the z operator is a shift operator, Equation (29) is shown in cyclic form of Equation (30).

$$u(n+2) = \left(K_P + \frac{T}{2}K_I + \frac{K_D}{T}\right)e(n+2) + \left(\frac{T}{2}K_I - K_P - \frac{2K_D}{T}\right)e(n+1) + \frac{K_D}{T}e(n) + u(n+1) \tag{30}$$

Here, e(n)is the nth sample data of the error signal.

If $K_P = 1$, $K_D = 0.005$, $K_I = 2$ and the sampling time T is assumed to be 0.001s, Equation (30) can be represented as Equation (31).

$$u(n+2) = 6.001e(n+2) - 10.999e(n+1) + 5e(n) + u(n+1) \tag{31}$$

If the PID controller can be shown in cyclic form, this can be conveniently programmed using C. Even here, the sampling time of the overall program has to be programmed equal to the sampling time of the PID controller.

3.3.2. *Parameter identification*

With regard to the whole system including the PID controller, the transfer function from the reference input r to the displacement x of the levitating object is as shown in Equation (32).

$$G_T = \frac{b_{m3}s^3 + b_{m2}s^2 + b_{m1}s + b_{m0}}{a_{m5}s^5 + a_{m4}s^4 + a_{m3}s^3 + a_{m2}s^2 + a_{m1}s + a_{m0}} \tag{32}$$

Here,

$$
\begin{aligned}
a_{m5} &= R_I R_F C_f L m \\
a_{m4} &= R_I R_F C_f m(R + R_s) + R_I R_d C_f m \\
a_{m3} &= R_I m - R_I R_F C_f(bcR_s + aL) - bR_s R_F R_d C_f K_D G_s \\
a_{m2} &= -a(R_I R_F C_f(R + R_s) + R_I R_d C_f) - bR_s R_F(K_D + R_d C_f K_P)G_s \\
a_{m1} &= -aR_I - bR_s R_F(K_P + R_d C_f K_I)G_s \\
a_{m0} &= -bR_s R_F K_I G_s \\
b_{m3} &= -bR_s R_F R_d C_f K_D \\
b_{m2} &= -bR_s R_F(K_D + R_d C_f K_P) \\
b_{m1} &= -bR_s R_F(K_P + R_d C_f K_I) \\
b_{m0} &= -bR_s R_F K_I \\
a &= \frac{2k I_{ss}^2}{X_1^3} \\
b &= -\frac{2kI_{ss}}{X_1^2} \\
c &= \frac{2kI_{ss}}{X_1^2} \\
X_1 &= X_0 + W + x \\
X_0 &= \frac{l_m}{2\mu_s}
\end{aligned}
$$

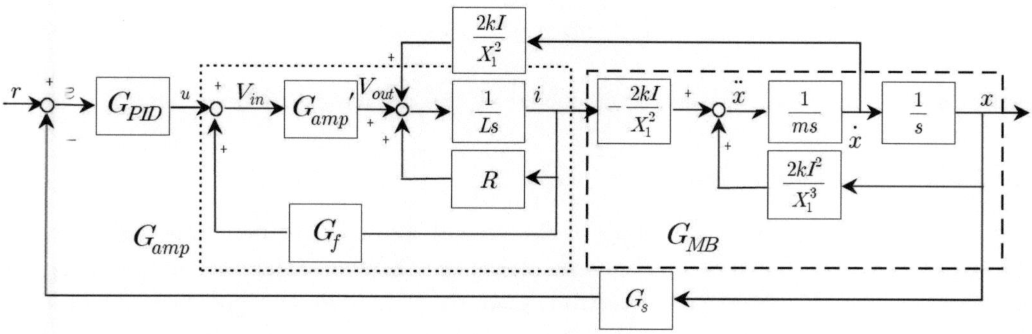

Figure 14. Block diagram of the magnetic bearing system including PID controller

Fig. 14 shows the block diagram of the whole system including the PID controller. To identify the unknown parameters using the genetic algorithm, the parameters difficult to measure has to be defined from the system transfer function and the search range of each parameter has to be defined. Using the error between the step response of the whole system and the data value obtained from experimentation, the objective function of the genetic algorithm is defined and the parameters of the genetic algorithm are defined.

Appendix 2 is the genetic algorithm program that allows the calculation of the unknown parameters from the above processes. At this point, the program part that is the same with the program to solve the amplifier peripheral circuit was excluded. Fig. 15 and Fig. 16 shows the graphs of the output results of the magnetic bearing system identification using RCGA similarly with Appendix 2.

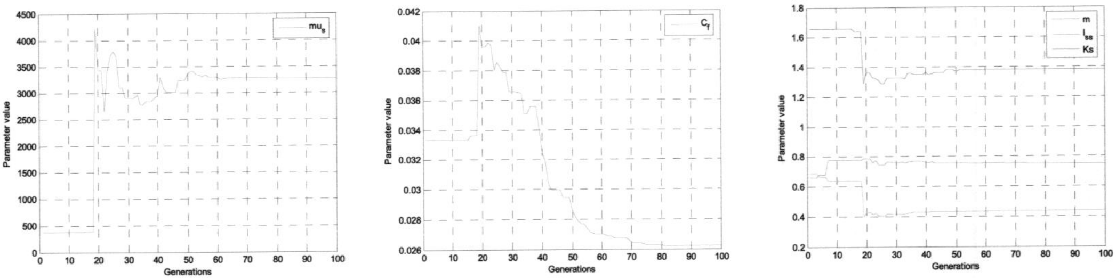

Figure 15. Trends of optimum parameters for each generation

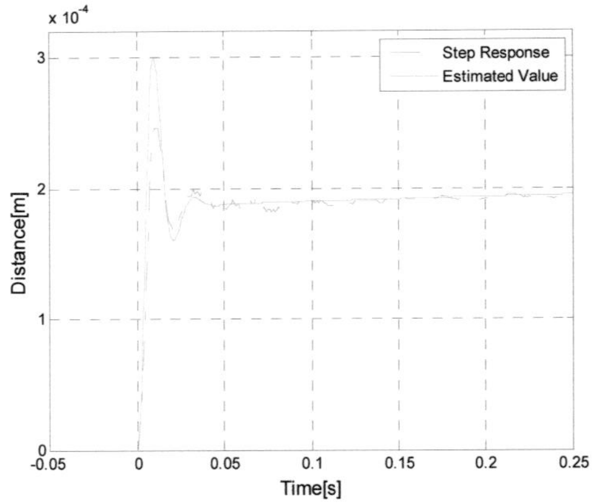

Figure 16. Experiment data comparison with the system obtained through RCGA results

3.4. Control signal division

The system of Fig. 14 is a modeling of the case where the magnetic bearing system is assumed to a horizontally symmetric based on the center point so that the levitating object is supported using an upper based electromagnet about the left or right parts. In order to properly levitate the levitating object of this system, a control signal equal to that of Fig. 14 has to be implemented and consistently supplied to the left and right magnetic bearing.

Also, to divide the control signal that supports the levitating object using only the upper electromagnet like that of Fig. 14 into the upper and lower electromagnet, the current i_{up} flowing in the upper electromagnet coil has to include the attractive force caused by the current i_{down} flowing in the lower electromagnet coil, where Equation (33) has to be followed for the design.

$$i_{up} = I_{ss} + i_{upo} + \alpha \tag{33}$$

Here, i_{up0} is the control current necessary to support the levitating object with only the upper electromagnet and α is the current corresponding to the attractive force caused by the current flowing in the lower electromagnet coil.

At this point, the magnetic bearing system has a form symmetric about each axis. Therefore, when assuming the axis parallel to the direction vertical from the Earth as the x-axis, the attractive force control of the y- and z-axes is the same as the x-axis control case including the neutral state and excluding the control current I_{ss}.

Fig. 17 is step response test result that is an example. When the program is work, levitated object is attracted by upper electric magnet. After that, the control signal is separated between upper and lower electric magnet. In the step response test, the disturbance mass is 150g.

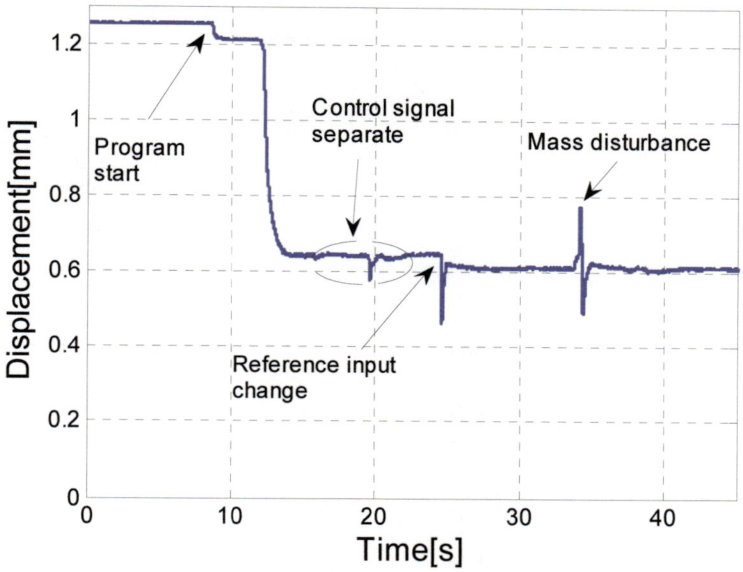

Figure 17. Step response test result for levitating system

3.5. Levitating object and equation of rotational motion

Fig. 18 shows a schematic of a magnetic bearing system taking into consideration rotational motion.

When the levitating object undergoes rotational motion, the torque caused by the rotational motion about the x-z plane is as shown in Equation (34).

$$J\ddot{\theta}_y - J_p\omega\dot{\theta}_x = -lf_{xl} + lf_{xr} \tag{34}$$

Here, J is the moment of inertia of the levitating object about the y-axis and J_p is the moment of inertia of the rotating levitating object about the x-axis. From Fig. 16, Equations (35) and (36) can be obtained.

$$\sin\theta_x = \frac{\Delta x}{l} \approx \theta_x \tag{35}$$

$$\sin\theta_y = \frac{\Delta y}{l} \approx \theta_y \tag{36}$$

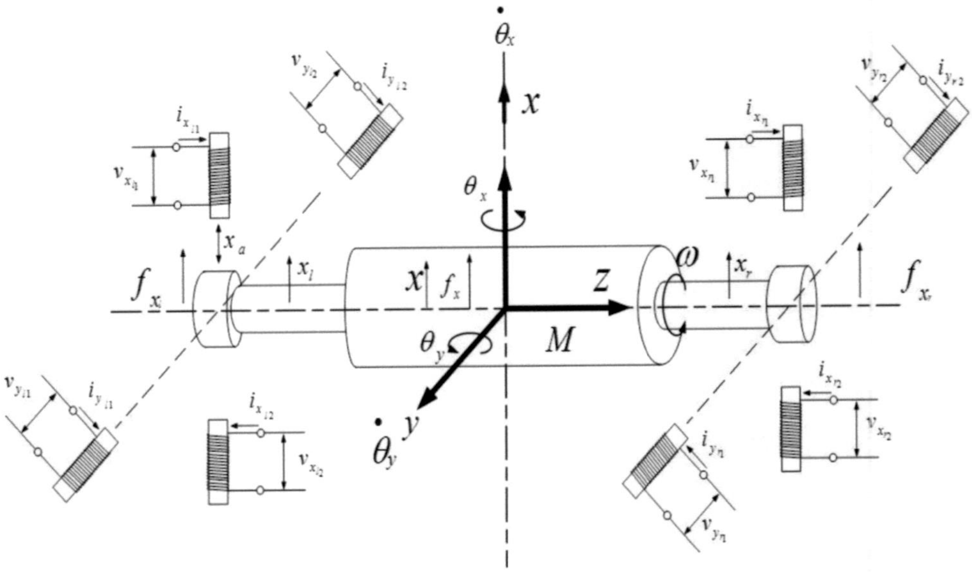

Figure 18. Magnetic bearing system taking into consideration rotational motion

When the levitating object undergoes rotational motion, the torque caused by the rotational motion about the y-z plane is as shown in Equation (37).

$$J\ddot{\theta}_x - J_p\omega\dot{\theta}_y = -lf_{yl} + lf_{yr} \tag{37}$$

In order to control the rotating levitating object, application of a multi-variable controller using state-space expression is necessary.

4. Conclusion

In this chapter, the detailed control laws to control the magnetic bearing was not included, and the method to designing a magnetic bearing system, obtaining a mathematical model, and the preparations necessary for control were explored with the aim of implementing a magnetic bearing system to apply various controllers.

Appendix

1. Attractive Force Calculation of the Electromagnet Using Probable Flux Paths Method
2. RCGA Program for Amplifier Peripheral Circuit Design
3. RCGA Program for Magnetic Bearing System Identification

1. Attractive force calculation of the electromagnet using probable flux paths method

It is assumed that the levitating object is supported by the electromagnet. Here, the gap between the electromagnet and the levitating object is 0.6mm, the number of coil winding to the electromagnet is 400turns, and the current flowing in the coil is 1A.

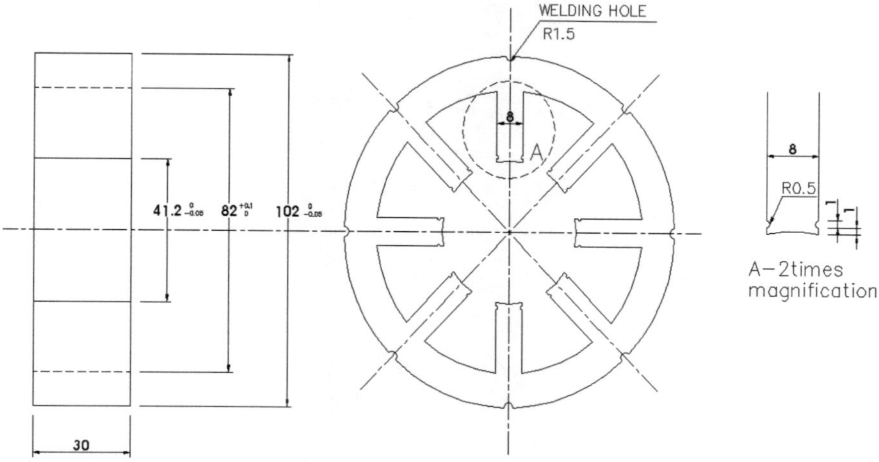

Figure 19. Electromagnet core drawing

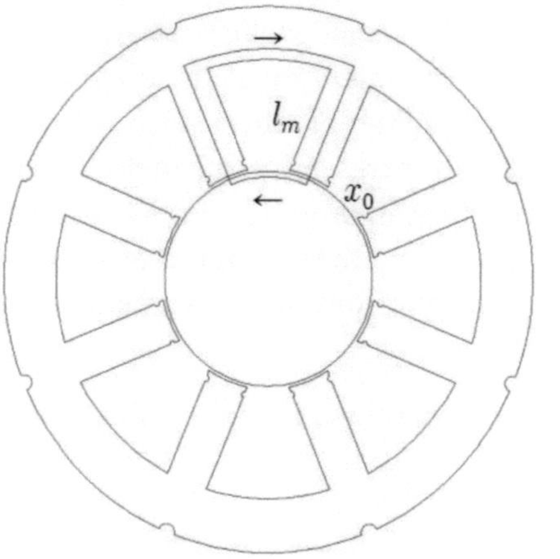

Figure 20. Magnetic circuit of the electromagnet

Fig. 19 shows the drawing of the electromagnet core. Fig. 20 shows the magnetic circuit that satisfies the Probable Flux Paths Method in the core of Fig. 19. The attractive force of the electromagnet is as shown in Equation (38).

$$F_m = \frac{\mu_0 N^2 I_m^2 S}{\left(\frac{l_m}{\mu_s} + 2x_0\right)^2} \tag{38}$$

Here, the length of the magnetic path is as shown in Equation (39) according to Fig. 2.

$$l_m = \frac{\pi d_1}{8} + \frac{\pi d_2}{8} + 2h = \left(\frac{\pi(82+41.2)}{8} + 2 \times 40\right) \times 10^{-3}[m] \tag{39}$$

When assuming the material of the electromagnet core as silicon steel plate($\mu_s = 3000$), the attractive force of this electromagnet is calculated using Equation (40).

$$F_m = \frac{4\pi\times10^{-7}\times400^2\times1^2\times2\times30\times10^{-3}\times0.8\times10^{-3}}{\left(\left(\frac{\pi(82+41.2)}{8\times3000}+2\times40+2\times0.6\right)\times10^{-3}\right)^2} = 6.2485[N] \tag{40}$$

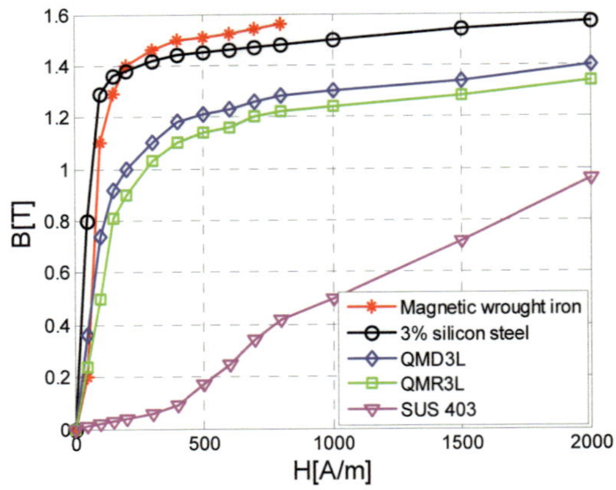

Figure 21. B-H curve of a magnetic substance

Fig. 21 shows the B-H curve of generic magnetic substances. As can be observed in Fig. 21, the magnetic flux density is concentrated about the magnetomotive force above a certain level for magnetic substances, implying that the attractive force does not increase when the current of the coil is increased to increase the magnetomotive force as the magnetic flux density does not increase. Therefore, in order to identify the maximum attractive force of an electromagnet, a process in identifying the maximum attractive force that can be used from the saturated magnetic flux density of the electromagnet material or the maximum current that can be bled in the coil is necessary.

The relationship between the current flowing in the coil and the magnetic flux density is as shown in Equation (41).

$$B = \frac{\Phi}{S} = \frac{\mu_0 N I_m}{\frac{l_m}{\mu_s}+2x_0} \tag{41}$$

From this, the maximum current I_{max} from the saturated magnetic flux density B_{max} can be solved as shown in Equation (42). However, the saturated magnetic flux density of silicon steel is approximately 1.5T.

$$I_{max} = \frac{\left(\frac{l_m}{\mu_s}+2x_0\right)B_{max}}{\mu_0 N}$$

$$I_{max} = \frac{\left(\frac{\pi(82+41.2)}{8\times3000}+2\times40+2\times0.6\right)\times10^{-3}\times1.5}{4\pi\times10^{-7}\times400} = 3.7087[A] \tag{42}$$

Here, the electromagnet maximum attractive force when the gap is 0.6mm is as shown in Equation (43).

$$F_{max} = \frac{4\pi\times10^{-7}\times400^2\times3.7087^2\times2\times30\times10^{-3}\times0.8\times10^{-3}}{\left(\left(\frac{\pi(82+41.2)}{8\times3000}+2\times40+2\times0.6\right)\times10^{-3}\right)^2} = 85.9442[N] \quad (43)$$

2. RCGA program for amplifier peripheral circuit design

Fig. 22 shows the linear current amplifier circuit. Considering that the voltage in Fig. 22 is V'_{fb}, the current can be expressed as Equation (44) and Equation (45) can be rearranged to obtain Equation (45).

$$i_1 = i_2 + i_3 \quad (44)$$

$$\frac{V_{IN}-V_{fb}}{R_I} = \frac{V_{fb}-V'_{fb}}{R_F} + \frac{V_{fb}-V_o}{Z_F} \quad (45)$$

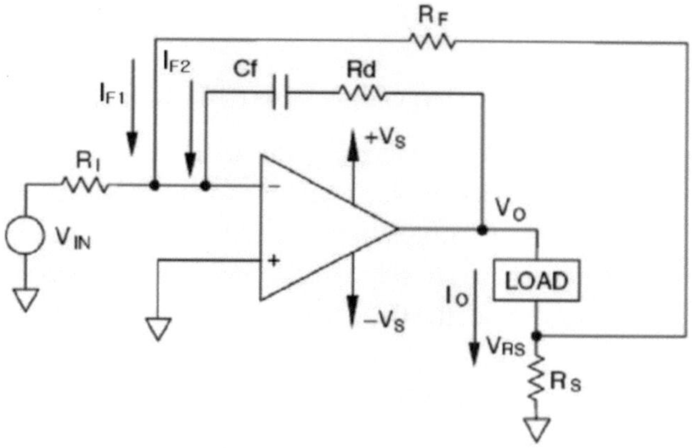

Figure 22. Linear current amplifier circuit

Also, considering that the voltage in Fig. 22 is V'_{fb}, the current can be expressed as Equation (46), and Equation (46) can be rearranged to obtain Equation (47).

$$i_2 + I_o = i_4 \quad (46)$$

$$\frac{V_{IN}-V'_{fb}}{R_F} + I_o = \frac{V'_{fb}-0}{R_s} \quad (47)$$

Equation (48) is obtained from the open-loop gain of the operational amplifier and the relationship of Equation (49) is obtained from the current flowing in the load.

$$V_{fb} = -\frac{V_o}{A} \quad (48)$$

$$V'_{fb} = V_o - I_o Z_L \quad (49)$$

When the transfer function is found from the relationships of Equations (45), (47), and (48), it is as shown in Equation (50).

$$\frac{I_o}{V_{IN}} = \frac{n'}{d'} \tag{50}$$

Here,

$$n' = -n_1 s - n_0$$
$$n_1 = \left(\frac{R_s}{A} + R_s + R_F\right) R_F R_d C_f$$
$$n_0 = \left(\frac{R_s}{A} + R_s + R_F\right) R_F$$
$$d' = (a_1 a_2 - a_3)s^2 + (a_1 b_1 + a_2 b_2 - b_3)s + b_1 b_2 - c_1$$
$$a_1 = b_2 R_d C_f + \frac{R_I R_F}{A} C_f + R_I R_F C_f$$
$$a_2 = (R_s + R_F)L$$
$$a_3 = \left(\frac{R_s}{A} + R_s + R_F\right) R_I R_d C_f L$$
$$b_1 = R_s R + R_F R_s + R_F R$$
$$b_2 = \frac{R_F}{A} + \frac{R_I}{A} + R_I$$
$$b_3 = \left(\frac{R_s}{A} + R_s + R_F\right) R_I (R_d C_f R + L)$$
$$c_1 = \left(\frac{R_s}{A} + R_s + R_F\right) R_I R$$

When A is assumed to be sufficiently large, Equations (50) and (23) are equivalent.

Before selecting the current amplifier circuit part values, the parts to find the value of are first identified. The parts to determine first are the amplifier (-)input resistance R_I and the feedback line resistance R_F. These resistances are parameters that determine the amplification ratio of the amplifier output current and are resistances to transfer the voltage signal. R_I and R_F has the relationship of Equation (51) from the DC gain of the closed circuit transfer function.

$$R_F = -R_I R_s \frac{I_o}{V_{IN}} \tag{51}$$

The ratio of the current amplifier input voltage and output current to be designed is assumed to be 1 : 1 and accordingly, R_F has the same relationship as shown in Equation (52).

$$R_F = -R_I R_s \tag{52}$$

The current amplifier amplification ratio A, current limiting resistance R_s, load inductance L and resistance R values are given as organized in Table. 1. Therefore, the variables of genetic algorithm to determine are limited to the resistance R_I which determines the amplifier output current amplification ratio, resistance R_d which determines the dynamic characteristics of the linear amplifier circuit, and condenser C_f.

Parameter	Value
A	$10^{107/20}$
L	9.2mH
R	6.2Ω
R_s	2Ω

Table 1. The known parameter's value at the current amplifier

In the case of resistance R_I which determines the amplifier output current amplification ratio, it has to be sought after in the range of kΩ. In comparison, resistance R_d, which determines the dynamic characteristics of the linear amplifier circuit, has to be sought after in a wide range. It is desirable to use a value for the condenser C_f which determines the dynamic characteristics of linear amplifier circuit in the nF to μF range in consideration of the dynamic characteristics of the current amplifier.

Therefore, the search ranges are determined as shown in Equations (53), (54), and (55).

$$8000 \leq R_I \leq 11000 \tag{53}$$

$$10^{-1} \leq R_d \leq 10^6 \tag{54}$$

$$10^{-10} \leq C_f \leq 10^{-7} \tag{55}$$

The transfer function with a desired output is defined as shown in Equation (56) with the same form as the transfer function of a current amplifier circuit. This is so that the desired output can be achieved with the combination of part values of the current amplifier circuit.

$$G_r(s) = -\frac{e_1 s + e_0}{s^2 + d_1 s + d_0} \tag{56}$$

Equation (57) was obtained through trial and error in an effort to obtain a step response rise time below 0.001s and the percentage overshoot below 5%. Fig. 23 shows the step response of the transfer function of Equation (57).

$$G_r(s) = -\frac{3000s + 210000}{s^2 + 3600s + 210000} \tag{57}$$

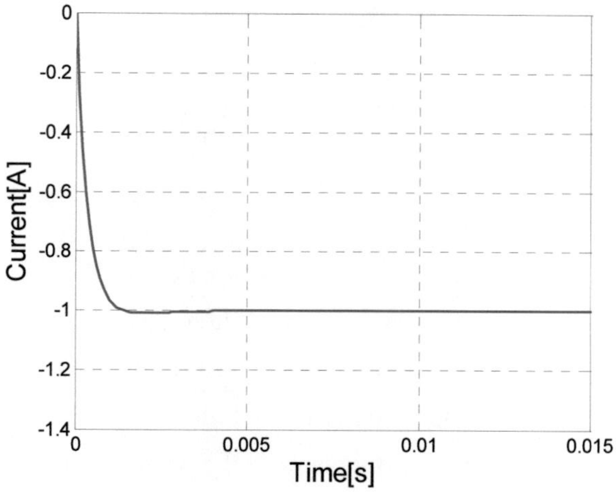

Figure 23. The desired step response

The objective function to apply genetic algorithm is Equation (58).

$$F_{obj} = \int e(t)^2 dt \tag{58}$$

Here, $e(t) = y_r(t) - y_{out}(t)$, $y_r(t)$ is the step response of the system satisfying the required time response characteristics and $y_{out}(t)$ is the current amplifier circuit transfer function step response obtained from the selected chromosome.

Table 2 shows the parameter values necessary to implement the real coded genetic algorithm.

The real coded genetic algorithm problem was implemented through matlab and the solution could be found by executing the rcga.m file. The names of files linked to rcga.m and their functions are shown in Table 3. Each script is as follows.

Parameter	Parameter value	Genetic operator
Population	50	Generate initial population
Max. generation	200	Generate initial population
Chromosome length	3	Generate initial population
Crossover probability	0.9	Modified simple crossover
Mutation probability	0.1	Dynamic mutation
eta	1.7	Scale fitting

Table 2. The parameter's value of the RCGA for the current amplifier design

file name (*.m)	function
rcga	Find optimal value on object function with RCGA
rInitPa	Define program variable for RCGA
rInitPop	Initialize the population
EvalObj	Evaluate the object function on the population for the reproduction
rGradSel	Reproduction operator with a gradient like selection method
rMsXover	Crossover operator with a modified simple crossover method
rDynaMut	Mutation operator with a dynamic mutation method
rElitism	Let to survive the best chromosome at the present generation to next genetation
ScaleFit	To improve the reproduction operator's efficience
rStatPop	Memorize the poplation's state for each generation

Table 3. Program file list for the execution of the real coded genetic algorithm

```
% rcga.m
% The RCGA implements a real coded genetic algorithm for finding the
% component value in the current amp. circuit
%
%  Encoding:
%     - Real
%
%  Genetic operators:
%     - Gradient-like selection
%     - Modified simple crossover
%     - Dynamic mutation
%
%  Other strategies:
%     - Elitism
%     - scaling window scheme(Ws=1)
%
% Remarks:
%
% Copyright (c) 2000 by Prof. Gang-Gyoo Jin, Korea Maritime
University
% Revision 0.9  2003/4/17
% Edit by Hwanghun Jeong, CME PKNU

clear;

% initializes the generation counter
gen= 1;

% initializes the parameters of a RCGA
[rseed,maxmin,maxgen,popsize,lchrom,pcross,pmutat,xlb,xub,etha,Ev]=
rInitPa;

% creates a polulation randomly
pop= rInitPop(rseed,popsize,lchrom,xlb,xub);

% calculates the objective function value
objfunc= EvalObj_new3(pop,lchrom,popsize);

% calculates gam
if(maxmin == 1)
  gam= min(objfunc);
else
  gam= min(-objfunc);
end

% calculates fitness using the scaling window scheme
fitness= ScaleFit(objfunc,popsize,gam,maxmin);

% computes statistics
[chrombest,objbest,fitbest,objave,gam]=
rStatPop(pop,objfunc,fitness,maxmin);
```

```
% builds a matrix storage for plotting line graphs
 stats(gen,:)=[gen objbest objave chrombest];

for gen= 2:maxgen

% prints the current generation
    fprintf('gen= %d (%d)\n',gen,maxgen-gen);

% applies reproduction
    pop=
rGradSel(pop,popsize,lchrom,fitness,chrombest,fitbest,xlb,xub,etha);
% Gradient-like selection

% applies crossover
    [pop,nxover]= rMsXover(pop,popsize,lchrom,pcross); % modified
simple crossover

% applies mutation
    [pop,nmutat]=
rDynaMut(pop,popsize,lchrom,pmutat,xlb,xub,gen,maxgen); %dynamic
mutation

% calculates the objective function value
  objfunc= EvalObj_new3(pop,lchrom,popsize);

% applies modified Elitism
[pop,objfunc]= rElitism(pop,objfunc,chrombest,objbest,maxmin);

% applies the scaling window scheme
    fitness= ScaleFit(objfunc,popsize,gam,maxmin);

% computes statistics
    [chrombest,objbest,fitbest,objave,gam]=
rStatPop(pop,objfunc,fitness,maxmin);

% builds a matrix storage for plotting line graphs
    stats(gen,:)=[gen objbest objave chrombest];

end

figure(1)
% plots the best and average objective function values
subplot(2,1,1)
plot(stats(:,1),stats(:,2))
xlabel('Generation'),ylabel('object function')

% plots the variables of the best chromosome
subplot(2,1,2)
plot(stats(:,1),stats(:,4),'-',stats(:,1),stats(:,5),'--
',stats(:,1),stats(:,6),'--')
xlabel('Generation'),ylabel('control parameter')
```

```
legend('R1','Rd','Cf')


figure(2)

Rs = 2; R = 6.2; A = 10^(107/20); L = 9.2 * 10^-3;

i=1;
R1 = chrombest(1,1) ;
Rf = Rs * R1;
var(i,1) = chrombest(1,2);
var(i,2) = chrombest(1,3);

a = -(Rs/A +Rs +Rf)*Rf*var(i,1)*var(i,2);
b = -(Rs/A +Rs +Rf)*Rf;
c = ((Rf/A + R1/A +R1)*var(i,1)*var(i,2) + R1*Rf*var(i,2)/A +
R1*Rf*var(i,2))*(Rs + Rf)*L - (Rs/A + Rs
+Rf)*R1*var(i,1)*var(i,2)*L;
d1 = ((Rf/A + R1/A +R1)*var(i,1)*var(i,2) + R1*Rf*var(i,2)/A +
R1*Rf*var(i,2))*(Rs*R + Rf*Rs + Rf*R) + (Rs + Rf)*L*(Rf/A + R1/A
+R1) - (Rs/A + Rs +Rf)*R1*(var(i,1)*var(i,2)*R + L);
e = (Rf/A + R1/A +R1)*(Rs*R + Rf*Rs + Rf*R) - (Rs/A + Rs +Rf)*R1*R;

n=[a b];
d=[c d1 e];
h= 0.0001; wdata = 150; t=0:h:wdata*h;

yn=[3000 2100000];
yd=[1 3600 2100000];
r2=step(-yn,yd,t);
r1=step(n,d,t);

plot(t,r1,'-',t,r2,'--')
legend('yout','yr')
xlabel('Time[s]'),ylabel('Current[A]')


% rInitPa.m
% The RINITPA function initializes the parameters of a RCGA
%
% Output:
%    rseed- random seed
%    maxmin= -1 for minimization, 1 for maximization
%    maxgen-    maximum generation
%    popsize- population size(must be an even integer)
%    lchrom- chromosome length
%    pcross- crossover probability
%    pmutat- mutation probability
%    xlb- lower bound  of variables
%    xub- upper bound of variables
%    etha- parameter of the selection operator
```

```
%
% Copyright (c) 2000 by Prof. Gang-Gyoo Jin, Korea Maritime
University
% Revision 0.9  2003/4/17
% Edit by Hwanghun Jeong, CME PKNU

function
[rseed,maxmin,maxgen,popsize,lchrom,pcross,pmutat,xlb,xub,etha,Ev]=
rInitPa

rseed=      8512;
maxmin=       -1;              % -1 for minimization
maxgen=    200;
popsize=     100;               % popsize should be even
lchrom=        3;
etha=        1.7;
pcross=      0.9;
pmutat=     0.1;
xlb(1,1) = 8000;
xlb(1,2) = 0.1;
xlb(1,3) = 1*10^-10;

xub(1,1) = 11000;
xub(1,2) = 1000000;
xub(1,3) = 1*10^-7;

Ev=0;

if(rem(popsize, 2) ~= 0) % do not move
    popsize= popsize + 1;
end


% rInitPop.m
%
% The RINITPOP function creates an initial population
%
% Input:
%    rseed- random seed
%    popsize- population size
%    lchrom- chromosome length
%    xub- upper bound for variables, vector
%    xlb- lower bound for variables, vector
% Output:
%    pop- population
%
% Copyright (c) 2000 by Prof. Gang-Gyoo Jin, Korea Maritime
University
% Revision 0.9  2003/4/17
% Edit by Hwanghun Jeong, CME PKNU
```

```
function pop= rInitPop(rseed,popsize,lchrom,xlb,xub)

rand('seed',rseed);
pop= zeros(popsize,lchrom);
for i=1:popsize
  pop(i,:)= (xub-xlb).*rand(1,lchrom)+xlb;
end

% EvalObj_new3.m
%
% The EVALOBJ function evaluates the objective function value
%
% Input:
%    var- variables, matrix
%    npara- number of the variables
%    popsize- population size
% Output:
%    objfunc- objective function value, vector
%
% Copyright (c) 2000 by Prof. Gang-Gyoo Jin, Korea Maritime
University
% Revision 0.9  2003/4/17
% Edit by Hwanghun Jeong, CME PKNU

function objfunc= EvalObj_new3(var,npara,popsize);

Rs = 2; R = 6.2; A = 10^(107/20); L = 9.2 * 10^-3;

for i= 1:popsize
objfunc(i)=0; oldobj=0;
R1 = var(i,1);
Rf = Rs * R1;
a = -(Rs/A +Rs +Rf)*Rf*var(i,2)*var(i,3);
b = -(Rs/A +Rs +Rf)*Rf;
c = ((Rf/A + R1/A +R1)*var(i,2)*var(i,3) + R1*Rf*var(i,3)/A +
R1*Rf*var(i,3))*(Rs + Rf)*L - (Rs/A + Rs
+Rf)*R1*var(i,2)*var(i,3)*L;
d1 = ((Rf/A + R1/A +R1)*var(i,2)*var(i,3) + R1*Rf*var(i,3)/A +
R1*Rf*var(i,3))*(Rs*R + Rf*Rs + Rf*R) + (Rs + Rf)*L*(Rf/A + R1/A
+R1) - (Rs/A + Rs +Rf)*R1*(var(i,2)*var(i,3)*R + L);
e = (Rf/A + R1/A +R1)*(Rs*R + Rf*Rs + Rf*R) - (Rs/A + Rs +Rf)*R1*R;

n=[a b];
d=[c d1 e];

h= 0.0001; wdata = 150; t=0:h:wdata*h;
yn=[-3000 -2100000];
yd=[1 3600 2100000];
yr = step(yn,yd,t);
resp = step(n,d,t);
err(:,1) =  resp(:,1) - yr(:,1);
```

```
for j= 1:wdata
    obj= err(j,1)^2;
    objfunc(i)= objfunc(i)+0.5*h*(obj+oldobj);
    oldobj= obj;
    end
end


% ScaleFit.m
%
% The SCALEFIT function converts objective function values into
fitness using
% the scaling window scheme(window size= 1)
%
% Input:
%    objfunc- objective function value, vector
%    popsize- population size
%    gam- minimun of objfunc or -objfunc in the previous population
%    maxmin= -1 for minimization, 1 for maximization
% Output:
%    fitness- scaled fitness, vector
%
% Copyright (c) 2000 by Prof. Gang-Gyoo Jin, Korea Maritime
University
% Revision 0.9  2003/4/17
% Edit by Hwanghun Jeong, CME PKNU

function fitness= ScaleFit(objfunc,popsize,gam,maxmin)

if(maxmin == 1)
  fitness= objfunc-gam;
else
  fitness= -objfunc-gam;
end
for i=1:popsize
  if(fitness(i) < 0)
    fitness(i)= 0;
  end
end


% rStatPop.m
%
% The RSTATPOP function calculates the statistics of a population
%
% Input:
%    pop- population, matrix
%    objfunc- objective function value, vector
%    fitness- fitness, vector
%    maxmin= -1 for minimization, 1 for maximization
```

```
% Output:
%     chrombest- best chromosome, vector
%     objbest- best objective function value
%     fitbest- fitness of the best chromesome
%     objave- average objective function value
%     gam- minimun of objfunc or -objfunc
%
% Copyright (c) 2000 by Prof. Gang-Gyoo Jin, Korea Maritime
University
% Revision 0.9  2003/4/17
% Edit by Hwanghun Jeong, CME PKNU

function [chrombest,objbest,fitbest,objave,gam]=
rStatPop(pop,objfunc, ...

fitness,maxmin)

if(maxmin == 1)
    [objbest, index]= max(objfunc);
    gam= min(objfunc);
else
    [objbest, index]= min(objfunc);
    gam= min(-objfunc);
end
chrombest= pop(index,:);
fitbest= fitness(index);
objave= mean(objfunc);


% rGradSel.m
%
% The RGRADSEL function performs gradient-like selection
%
% Input:
%     pop- population of chromosomes, matrix
%     popsize- population size
%     lchrom- chromosome length
%     fitness- fitness, vector
%     chrombest- best chromosome, vector
%     fitbest- fitness of the best chromesome
%     xlb- lower bound for variables, vector
%     xub- upper bound for variables, vector
%     etha- parameter of the selection operator
% Output:
%     newpop- mating pool, matrix
%                                                                  %
% Copyright (c) 2000 by Prof. Gang-Gyoo Jin, Korea Maritime
University
% Revision 0.9  2003/4/17
% Edit by Hwanghun Jeong, CME PKNU
```

```
function newpop=
rGradSel(pop,popsize,lchrom,fitness,chrombest,fitbest,xlb, ...

xub,etha)
if(fitbest > 0)
    for i= 1:popsize
      etha1= etha;
        normfit= 1-fitness(i)/fitbest;
        pass= 0;
        while(pass == 0)
            pass= 1;
            for j= 1:lchrom
                newpop(i,j)= pop(i,j)+etha1*normfit*(chrombest(j)-
pop(i,j));
                if(newpop(i,j) < xlb(j) | newpop(i,j) > xub(j))
                    etha1= etha1*0.8;
                    pass= 0;
                    break;
                end
            end
        end
    end

else
    for i= 1:popsize
        k= Pickup(popsize);
        newpop(i,:)= pop(k,:);
    end
end

% Pickup.m
%
% The PICKUP function picks up an integer random number between 1
and num
%
% Input:
%    num- integer number greater than or equal to 1
% Output:
%    rnum- random number between 1 and num
%
% Copyright (c) 2000 by Prof. Gang-Gyoo Jin, Korea Maritime
University
% Revision 0.9  2003/4/17
% Edit by Hwanghun Jeong, CME PKNU

function rnum= Pickup(num)

if min(num) < 1
  disp('num is less than one !')
  return;
end
```

```
fr= rand(size(num));
rnum= floor(fr.*num)+1;


% rMsXover.m
%
% The RMSXOVER function performs modified simple crossover
%

% Input:
%    pop- population of chromosomes, matrix
%    popsize- population size
%    lchrom- chromosome length
%    pcross- crossover probability
% Output:
%    pop- mated population, matrix
%    nxover- number of times crossover was performed
%
% Copyright (c) 2000 by Prof. Gang-Gyoo Jin, Korea Maritime
University
% Revision 0.9  2003/4/17
% Edit by Hwanghun Jeong, CME PKNU

function [pop,nxover]= rMsXover(pop,popsize,lchrom,pcross)

nxover= 0;
halfpop= floor(popsize/2);
for i= 1:halfpop
    if (rand <= pcross)
        nxover= nxover+1;
       mate1= 2*i-1;
       mate2= 2*i;
        xpoint= Pickup(lchrom-1);
        lam= rand;
        temp= lam*pop(mate2,xpoint)+(1-lam)*pop(mate1,xpoint);
        lam= rand;
        pop(mate2,xpoint)= lam*pop(mate1,xpoint)+(1-
lam)*pop(mate2,xpoint);
        pop(mate1,xpoint)= temp;

        temp= pop(mate1,xpoint+1:lchrom);
        pop(mate1,xpoint+1:lchrom)= pop(mate2,xpoint+1:lchrom);
        pop(mate2,xpoint+1:lchrom)= temp;
    end
end


% rDynaMut.m
%
% The RDYNAMUT function performs dynamic mutation
%
```

```
% Input:
%    pop- population of chromosomes, matrix
%    popsize- population size
%    lchrom- chromosome length
%    pmutat- mutation probability
%    xlb- lower bound  of variables
%    xub- upper bound of variables
% Output:
%    pop- mutated population, matrix
%    nmutat- number of times mutation was performed
%
% Copyright (c) 2000 by Prof. Gang-Gyoo Jin, Korea Maritime
University
% Revision 0.9  2003/4/17
% Edit by Hwanghun Jeong, CME PKNU

function [pop,nmutat]=
rDynaMut(pop,popsize,lchrom,pmutat,xlb,xub,gen,maxgen)

b= 5;
nmutat= 0;
for i= 1:popsize
    for j= 1:lchrom
        if (rand <= pmutat)
            nmutat= nmutat+1;
            r= rand;
            if(round(rand))
                pop(i,j)= pop(i,j)+(xub(j)-pop(i,j))*r*(1-
gen/maxgen)^b;
            else
                pop(i,j)= pop(i,j)-(pop(i,j)-xlb(j))*r*(1-
gen/maxgen)^b;
            end
        end
    end
end


% rElitism.m
%
% The RELITISM function performs elitism
%
% Input:
%    pop- population of chromosomes, matrix
%    objfunc- objective function value
%    chrombest- best chromosome, vector
%    objbest- best objective function value
%    maxmin= -1 for minimization, 1 for maximization
% Output:
%    pop- modified population of chromosomes, matrix
%    objfunc- modified objective function value
```

```
%
% Copyright (c) 2000 by Prof. Gang-Gyoo Jin, Korea Maritime
University
% Revision 0.9  2003/4/17
% Edit by Hwanghun Jeong, CME PKNU

function [pop,objfunc]=
rElitism(pop,objfunc,chrombest,objbest,maxmin)

if(maxmin==1)
     cobjbest= max(objfunc);
     if(cobjbest < objbest)
       [objworst, index]= min(objfunc);
       pop(index,:)= chrombest;
       objfunc(index)= objbest;
     end
else
     cobjbest= min(objfunc);
     if(cobjbest > objbest)
       [objworst, index]= max(objfunc);
       pop(index,:)= chrombest;
       objfunc(index)= objbest;
     end
end
```

3. RCGA program for magnetic bearing system identification

The transfer function from the reference input of the magnetic bearing system including the PID controller to the displacement of the levitating object is as shown in Equation (59).

The PID controller coefficients selected for the stabilization of the magnetic bearing system are $K_P = 1$, $K_D = 0.005$, $K_I = 2$ and the sampling time is 0.001s. If the PID controller is expressed in cyclic form to implement as a micro processor, it is as shown in Equation (59).

$$u(n+2) = 6.001e(n+2) - 10.999e(n+1) + 5e(n) + u(n+1) \qquad (59)$$

Fig. 25 is the step response that was obtained from the magnetic bearing system including the PID controller designed for stabilization. Specially, Fig. 25 is the step response of the displaced levitating object displacement x when the right side electromagnet reference input was modified from 0.4mm to 0.6mm where the left side electromagnet was fixed.

Fig. 60 shows the connection diagram of the magnetic bearing control system. The power of the system uses a DC power supply, displacement sensor amplifier, DSP, and AC220V power for the PC, and DC power is used for the current amplifier. The control system is connected to the magnetic bearing coil and displacement measurement sensor through a port. The delivered signal from the displacement sensor amplifier is compared to the reference input in the DSP and a control signal is generated, where the control signal generated in the DSP is provided to the linear current amplifier circuit to control the electromagnet. The signals occurring during control are stored in the independently installed PC through the DAQ board(PCI6010) for monitoring.

The MPU to implement the PID controller is TMS320C32, and a 12 bit A/D converter (MAX122) and 12 bit D/A converter (AD664) were used. Eddy current type sensor (AH-305) was used as the displacement measurement sensor for the feedback signal and an appropriate sensor amplifier (AS-440-01) was applied.

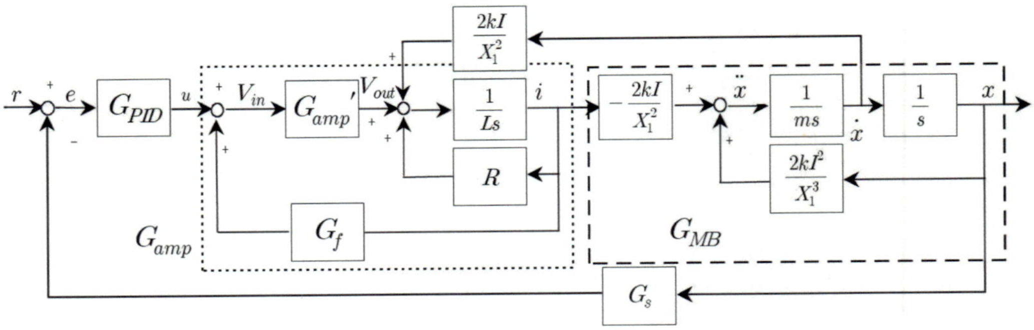

Figure 24. Magnetic bearing system including the PID controller

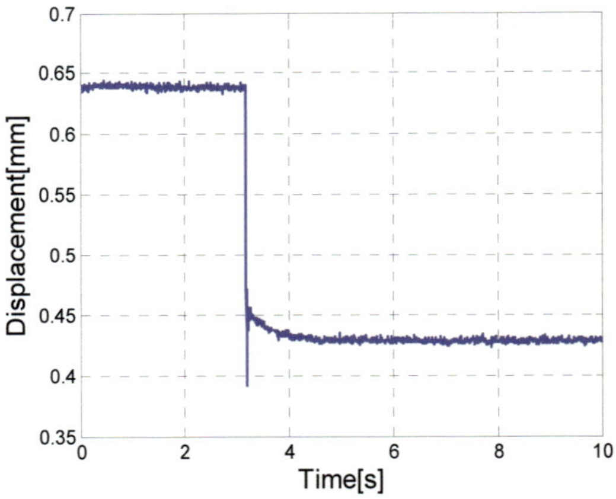

Figure 25. Step response of magnetic bearing system including the PID controller

Table 4 shows the parameter values already known regarding the magnetic bearing system. Therefore, the parameters that their values cannot be relatively exactly known in this system are the levitating object mass m, the current during normal state I_{SS}, coil inductance value L, relative permeability μ_s of the levitating object, and additional gain K_S for normal deviation calibration.

In the experiment for the identification of the magnetic bearing, the levitating object was supported using one side of the magnetic bearing. When supporting the levitating object with one side of the electromagnet, the levitating object becomes slanted so the vertical direction force that the electromagnet supports varies with the tilted angle of the levitating object and the impact force(mass m) of the levitating object on the electromagnet is difficult to measure. If the mass of the levitating object changes, the current I_{SS} at normal state depending on the mass also varies. Additionally, the coil inductance value L varies depending on the levitating object location within the electromagnet coil, thus, it is a parameter that is difficult to exactly measure. The relative permeability μ_s of the levitating object is also difficult to exactly obtain due to the uneven nature of the material, and the random gain to calibrate the normal deviation that occurs due to the mathematical model error is defined as K_S and is additionally included in the list of parameters to be identified.

Equation (60) shows the search ranges of the 5 unknown parameter values to be estimated by using the genetic algorithm. For each parameter, the search range was determined based on the actual experimented system and with the consideration of the physical characteristics.

$$
\begin{gathered}
0.5 \leq m \leq 0.9 \\
0 \leq I_{SS} \leq 0.7 \\
0 \leq L \leq 0.1 \\
0 \leq \mu_S \leq 10000 \\
0 \leq K_S \leq 2
\end{gathered}
\tag{60}
$$

Parameter	Value
Length of a path for magnetic flux	0.1711m
Displacement for levitate object(at steady state)	0.6mm
Cross section of armature	4.8×10^{-4} m^2
Number of coil turn for magnetic bearing	200 turn

Table 4 The table for the known parameters

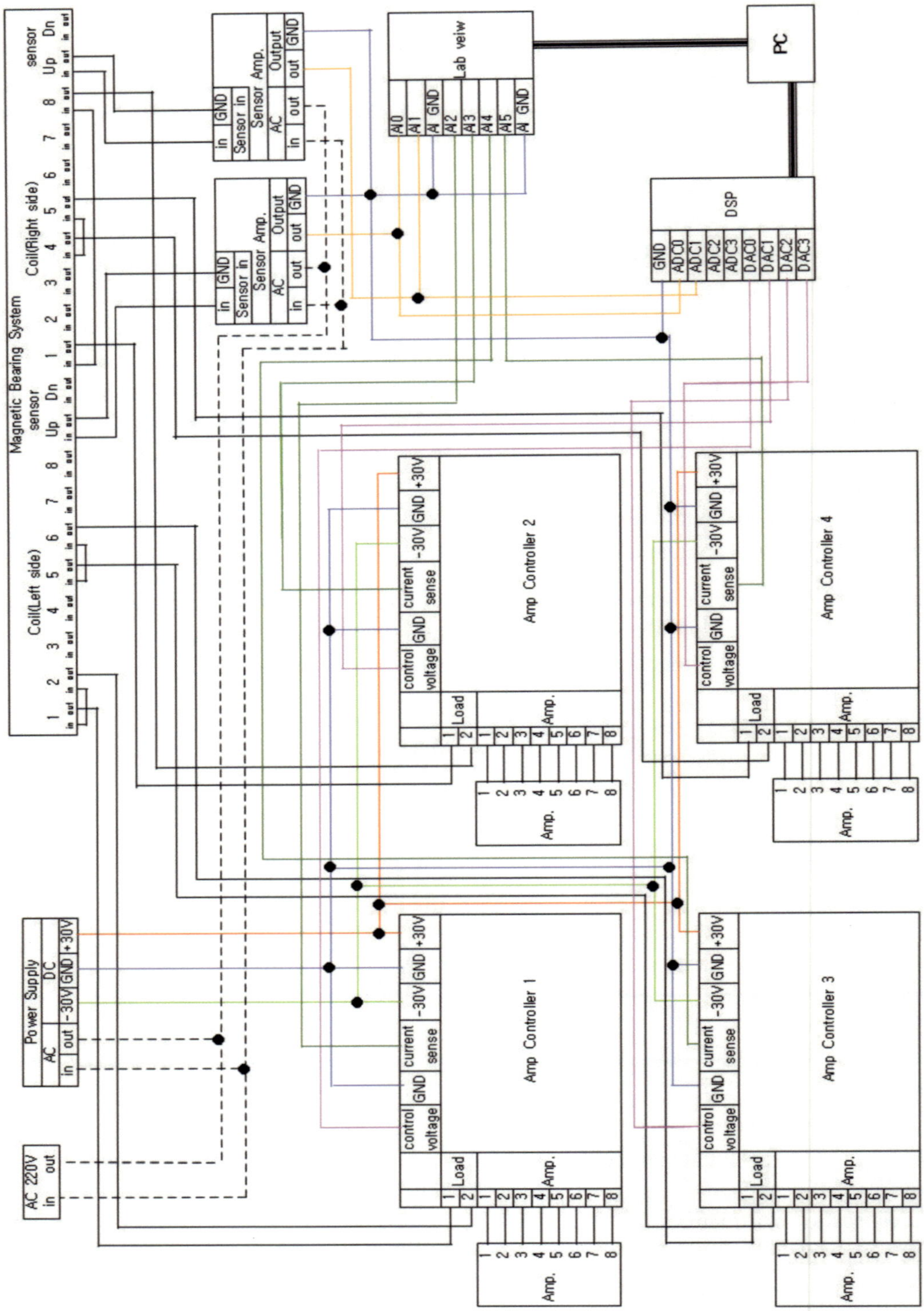

Figure 26. The connecting diagram for the magnetic bearing system

IAE(Integrated Absolute Error) as shown in Equation (61) was selected as the objective function to execute RCGA.

$$f_{obj} = \int_{t_0}^{t_f} |e(t)|\, dt \tag{61}$$

Here, e(t)is the difference between the magnetic bearing step response obtained experimentally and the magnetic bearing model step response obtained mathematically through the selected chromosome.

For the implement the real coded genetic algorithm, the program and the modified parts of Table 3 are shown in Table 6 and the scripts are as follows.

Parameter	Parameter Value	Genetic operator
Population	100	Generate initial population
Max. generation	100	Generate initial population
Chromosome length	5	Generate initial population
Crossover Probability	0.9	Modified simple crossover
Mutation Probability	0.1	Dynamic mutation
Eta	1.7	Scale fitting

Table 5. Shows the program parameters necessary to execute RCGA.

file name (*.m)	function
rcga	Find optimal value on object function with RCGA
rInitPa	Define program variable for RCGA
EvalObj	evaluate the object function on the population for the reproduction

Table 6. Modified program file list to execute real coded genetic algorithm

```
% rcga22.m
%
% The RCGA22 implements a real coded genetic algorithm for finding
system parameter in the MBS
%
%   Encoding:
%     - Real
%
%  Genetic operators:
%     - Gradient-like selection
%     - Modified simple crossover
%     - Dynamic mutation
%
%  Other strategies:
%     - Elitism
%     - scaling window scheme(Ws=1)
%
%  Remarks:
%
% Copyright (c) 2000 by Prof. Gang-Gyoo Jin, Korea Maritime
University
% Revision 0.9  2003/4/17
% Edit by Hwanghun Jeong, CME PKNU

clf;
clear;
Test_data = xlsread('a_pidR.xls');

% initializes the generation counter
gen= 1;

% initializes the parameters of a RCGA
[rseed,maxmin,maxgen,popsize,lchrom,pcross,pmutat,xlb,xub,etha]=
rInitPa22;

% creates a polulation randomly
pop= rInitPop(rseed,popsize,lchrom,xlb,xub);

% calculates the objective function value
objfunc= EvalObj22(pop,lchrom,popsize,Test_data);

% calculates gam
if(maxmin == 1)
  gam= min(objfunc);
else
  gam= min(-objfunc);
end

% calculates fitness using the scaling window scheme
fitness= ScaleFit(objfunc,popsize,gam,maxmin);
```

```
% computes statistics
[chrombest,objbest,fitbest,objave,gam]=
rStatPop(pop,objfunc,fitness,maxmin);

% builds a matrix storage for plotting line graphs
 stats(gen,:)=[gen objbest objave chrombest];

for gen= 2:maxgen

% prints the current generation
    fprintf('gen= %d (%d) %f\n',gen,maxgen-gen,objbest);

% applies reproduction
    pop=
rGradSel(pop,popsize,lchrom,fitness,chrombest,fitbest,xlb,xub,etha);

% Gradient-like selection
% applies crossover
    [pop,nxover]= rMsXover(pop,popsize,lchrom,pcross); % modified
simple crossover

% applies mutation
    [pop,nmutat]=
rDynaMut(pop,popsize,lchrom,pmutat,xlb,xub,gen,maxgen); %dynamic
mutation

% calculates the objective function value
    objfunc= EvalObj22(pop,lchrom,popsize,Test_data);

% applies Elitism
    [pop,objfunc]= rElitism(pop,objfunc,chrombest,objbest,maxmin);

% applies the scaling window scheme
    fitness= ScaleFit(objfunc,popsize,gam,maxmin);

% computes statistics
    [chrombest,objbest,fitbest,objave,gam]=
rStatPop(pop,objfunc,fitness,maxmin);

% builds a matrix storage for plotting line graphs
    stats(gen,:)=[gen objbest objave chrombest];
end

figure(1)
% plots the best and average objective function values
subplot(2,1,1)
plot(stats(:,1),stats(:,2:3))

% plots the variables of the best chromosome
subplot(2,1,2)
plot(stats(:,1),stats(:,4:lchrom+3))
axis([0 100 0 10000]);
```

```
figure(2)
Rd=273000;
Ri=10000*1.0045;
Rf=20000;
R=9;
A = 10^(107.7/20);
Gs=1000;
Kd=0.005;
Kp=1;
Ki=2;

Cf= 4.94*10^-9;
L_m=(1/8*pi*90+1/8*pi*40+2*60)*10^-3;
N=200;
mu_o=4*pi*10^-7;
X=0.0006;
S_a= 480*10^-6 *0.5;

m= chrombest(1); mu_s=chrombest(2);
I_ss=chrombest(3);L=chrombest(4);
pt1= chrombest(5);
Rs=pt1*2;

X_o=L_m/(2*mu_s);
X_1=X+X_o;

k=N^2*mu_o*S_a/4;

a=-2*k*I_ss^2/X_1^3;
b=-2*k*I_ss/X_1^2;
c=2*k*I_ss/X_1^2;

n=-Rs*b*Rf*[Rd*Cf*Kd (Rd*Cf*Kp + Kd) (Rd*Cf*Ki + Kp) Ki];

d01=Ri*Cf*Rf*m*L;
d02=Ri*Cf*Rf*m*(R + Rs) + Ri*m*Rd*Cf;
d03=-(Ri*Cf*Rf*(a*L + Rs*b*c) - Ri*m + Rd*Cf*Rs*Rf*b*Kd*Gs);
d04=-(Ri*Cf*Rf*a*(R + Rs) + Ri*a*Rd*Cf + (Rd*Cf*Rs*Rf*b*Kp +
Rs*Rf*b*Kd)*Gs);
d05=-(Ri*a + (Rd*Cf*Rs*Rf*b*Ki + Rs*Rf*b*Kp)*Gs);
d06=-Rs*Rf*b*Ki*Gs;

d=[d01 d02 d03 d04 d05 d06];

t_sample = 0:0.001:2.999;
    y=step(n,d,t_sample);
    plot(t_sample,Test_data(:,4)/1000,'-.',t_sample,0.21*y,'-')
    axis([-0.05 0.25 0 0.00032]);
legend('Step Response','Estimated Value')
xlabel('Time[s]'),ylabel('Distance[m]')
```

```
% rInitPa22.m
%
% The RINITPA22 function initializes the parameters of a RCGA
%
% Output:
%    rseed- random seed
%    maxmin= -1 for minimization, 1 for maximization
%    maxgen-    maximum generation
%    popsize- population size(must be an even integer)
%    lchrom- chromosome length
%    pcross- crossover probability
%    pmutat- mutation probability
%    xlb- lower bound  of variables
%    xub- upper bound of variables
%    etha- parameter of the selection operator
%
% Copyright (c) 2000 by Prof. Gang-Gyoo Jin, Korea Maritime
University
% Revision 0.9  2003/4/17
% Edit by Hwanghun Jeong, CME PKNU

function
[rseed,maxmin,maxgen,popsize,lchrom,pcross,pmutat,xlb,xub,etha]=
rInitPa20

rseed=       937;
%rseed=input('rseed= ');
maxmin=       -1;              % -1 for minimization
maxgen=  100;
popsize=     100;               % popsize should be even
lchrom=           5;
etha=          1.7;
pcross=        0.9;
pmutat=      0.1;
xlb=  0*ones(1,lchrom);
xub=  10*ones(1,lchrom);

%xlb(1,1)=0;
%xlb(1,2)=0;
%xlb(1,3)=0.6;
xlb(1,1)=0.5;
xlb(1,2)=0;
xlb(1,3)=0.7;
xlb(1,4)=0.025;

%xub(1,1)=0.5;
%xub(1,2)=1000;
%xub(1,3)=0.8;
xub(1,1)=0.9;
xub(1,2)=10000;
xub(1,3)=0.9;
```

```
xub(1,4)=0.095;
xub(1,5)=2;

%xub(1,2)=480*10^-6;
%xub(1,2)=1*10^-3;

%xub(1,1)=10000;

%xub(1,2)=10;
%xub(1,3)=3.5*10^5;
%xub(1,3)=1*10^-7;

if(rem(popsize, 2) ~= 0) % do not move
    popsize= popsize + 1;
end

% EvalObj22.m
%
% The EVALOBJ6 function evaluates a multivariable function
%
% Input:
%    x- variables, matrix
%    npara- number of the variables
%    popsize- population size
% Output:
%    objfunc- objective function value, vector
%
% Copyright (c) 2000 by Prof. Gang-Gyoo Jin, Korea Maritime
University
% Revision 0.9  2003/4/17
% Edit by Hwanghun Jeong, CME PKNU

function objfunc= EvalObj22(x,npara,popsize,Test_data);
Rd=273000;
Ri=10000*1.0045;
Rf=20000;
R=9;
A = 10^(107.7/20);
Gs=1000;
Kd=0.005;
Kp=1;
Ki=2;

Cf= 4.94*10^-9;
L_m=(1/8*pi*90+1/8*pi*40+2*60)*10^-3;
N=200;
mu_o=4*pi*10^-7;
X=0.0006;
S_a= 480*10^-6 *0.5;
```

```
for i= 1:popsize
m= x(i,1); mu_s=x(i,2); I_ss=x(i,3);L=x(i,4);
pt1=x(i,5);
Rs=pt1*2;

X_o=L_m/(2*mu_s);
X_1=X+X_o;

k=N^2*mu_o*S_a/4;

a=-2*k*I_ss^2/X_1^3;
b=-2*k*I_ss/X_1^2;
c=2*k*I_ss/X_1^2;

n=-Rs*b*Rf*[Rd*Cf*Kd (Rd*Cf*Kp + Kd) (Rd*Cf*Ki + Kp) Ki];

d01=Ri*Cf*Rf*m*L;
d02=Ri*Cf*Rf*m*(R + Rs) + Ri*m*Rd*Cf;
d03=-(Ri*Cf*Rf*(a*L + Rs*b*c) - Ri*m + Rd*Cf*Rs*Rf*b*Kd*Gs);
d04=-(Ri*Cf*Rf*a*(R + Rs) + Ri*a*Rd*Cf + (Rd*Cf*Rs*Rf*b*Kp +
Rs*Rf*b*Kd)*Gs);
d05=-(Ri*a + (Rd*Cf*Rs*Rf*b*Ki + Rs*Rf*b*Kp)*Gs);
d06=-Rs*Rf*b*Ki*Gs;

d=[d01 d02 d03 d04 d05 d06];

    t_sample = 0:0.001:2.999;
    y=step(n,d,t_sample);
    objfunc(i)=0;
    for j=1:3000
        obj=abs(Test_data(j,4)/1000-0.21*y(j,1));
        objfunc(i)= objfunc(i)+obj;
    end
end
```

Author details

Hwang Hun Jeong, So Nam Yun and Joo Ho Yang
Korea Institute of Machinery and Materials, Daejeon, Republic of Korea

Joo Ho Yang
PuKyong National University, Pusan, Republic of Korea

5. References

[1] Lichuan Li et al. A One-Axis-Controlled Magnetic Bearing and Its Performance. JSME international Journal. Series C 2003;46(2) 391-396.

[2] B. Polajzer et al. Decentralized PI/PD position control for active magnetic bearing . Electrical Engneering 2006; 53-59.

[3] Hector Martin Gutierrez and Paul I. Ro. Parametric Modeling and Control of Long-Range Actuator Using Magnetic Servo Levitation. IEEE Transaction on Megnetics 1998;(34)5 3689-3695.

[4] Susan Carlson-Skalak, Eric Maslen and Yong Teng. Magnetic Bearing Actuator Design using Genetic Algorthm. Journal of Engneering Design 1999;(10) 2 143-164.

[5] Kuan-Yu Chen et al. A Self-tunning fuzzy PID type controller design for unbalance compensation in a active magnetic bearing. Expert System with Application 36 2009; 8560-8570

[6] G. G. Jin. The Genetic Algorithm and the Apllications: Gyo Woo sa; 2000.

[7] Apex. Power Optional Amplifier, Apex, pp1-4. http://www.datasheet.co.kr/datasheet-html/P/A/1/PA12_ETC.pdf.html (accessed 13 July, 2012)

[8] Benjamin C. Kuo, The Digital Control Engineering: Hyoung Seol; 2003.